Lakhdar Meziani

Basic Structures in Mathematical Analysis

Lakhdar Meziani

Basic Structures in Mathematical Analysis

Mathematical Analysis

Noor Publishing

Imprint
Any brand names and product names mentioned in this book are subject to trademark, brand or patent protection and are trademarks or registered trademarks of their respective holders. The use of brand names, product names, common names, trade names, product descriptions etc. even without a particular marking in this work is in no way to be construed to mean that such names may be regarded as unrestricted in respect of trademark and brand protection legislation and could thus be used by anyone.

Cover image: www.ingimage.com

Publisher:
Noor Publishing
is a trademark of
International Book Market Service Ltd., member of OmniScriptum Publishing Group
17 Meldrum Street, Beau Bassin 71504, Mauritius

Printed at: see last page
ISBN: 978-620-0-77569-6

Basic Topics in Mathematical Analysis

Lakhdar Meziani

Batna University 05000 Batna Algeria.

Contents

Preface

The purpose of this book is to make available to the student some fundamentals of mathematical analysis. Specifically, it is intended to make such fundamentals available in a form that meets their need in many applications, like real analysis, integration, measure theory, and representation theory.

The principal point of view is to develop the basic structures of analysis, under which one can appropriately go on further in the domain of functionl analysis.

The book is intended to be essentially self contained and accessible to well-trained undergraduated students. Its prerequisites are elementary standards from basic algebra. In writing this book, we care about doing things as little abstract as possible. So, to make easy the access to the main concepts, each section of each chapter is illustrated by simple   examples and exercises, which are mostly applications to concrete problems.

References of treatises on the domain are given at the end. We hope that the book will reach the objectives assigned and especially will be useful to the reader.

Lakhdar Meziani

Introduction

In this introduction we give some landmarks to make the contents easy to use. Chapter 1 contains some basics of set theory: Binary relations, Real number system, and Cardinals are the main topics of the chapter.

Chapter 2 introduces topological structures and their morphisms, which are the continuous functions. Insistence is made on the construction processes of a topology on a given set. Essentially two processes are described: topology generated by means of a family of sets and topology generated by means of a family of functions. This in turn is applied to the construction of the product topology on the product of an arbitrary family of sets.

The notion of separation is of a particular importance in topology. Two types of separated topological spaces are considered: Hausdorff spaces and Normal spaces. The importance of the normal spaces is that one can define on them non trivial continuous functions by Urysohn Lemma.

Chapter 3 is intended to metric spaces. A metric on a set makes the construction of a topology easy, with the geometric notions of balls. The continuity of functions is expressed in terms of sequences limits. Separable spaces, Complete spaces and Baire spaces are among the most frequently used metric spaces in applications.

Chapter 4 contains basic facts about Compact spaces and Locally compact spaces. The properties of compact spaces are remarkable: for example any Compact space is normal, an a Metric compact space is separable. Locally compact spaces. are well adapted for special areas for example in integration theory. The prototype of locally compact space is the euclidean space $\mathbb{R}^n$.

Chapter 5 is a relatively detailled introduction to Banach spaces and their linear operators. Fundamental theorems used in functional analysis are given: the Open Mapping Theorem, the Closed Graph Theorem, the Banach-Steinhauss Theorem. Moreover duality and weak and weak* topologies in Banach spaces are considered in some details. Note also the outstanding Alaoglu theorem.

Chapter 6 is devoted to Hilbert spaces. The extension of the primitive form of the inner product and the notion of orthogonality to infinite dimensional vector spaces, undoubtly constitute the fondations of the hilbertian geometry.

Three important facts have to be emphasized in this context:

1. The existence of orthonormal base in any non trivial Hilbert space X, which allow to classify X as a space of type $l_2(A)$.

2. The orthogonal projection of a vector on a subspace of X and the approximation which follows.

3. The identification between a Hilbert space X and its dual X^* by Riesz Theorem.

Chapter 7 may be considered as an introduction to topological vector spaces with some basic properties. The notion of completeness, we will deal with, is of particular importance. This will be done through the concept of generalized Cauchy sequences.

Chapter 8 contains topological vector spaces of finite dimension.

Chapter 9 concerns the important space C(X) of continuous functions on a set X.

CHAPTER 1

Basics in Set Theory

1. Set Operations

In this chapter we give some basics from set theory and also some fundamental properties of the real number system.

Definition 1.1.1. Let X be a set. We denote by $\mathcal{P}(X)$ the power set of X that is, the family of all subsets of X. If $A, B \in \mathcal{P}(X)$ let us define:

$A \cup B = \{x \in X : \ x \in A \text{ or } x \in B\}$ (union of A and B)
$A \cap B = \{x \in X : \ x \in A \text{ and } x \in B\}$ (intersection of A and B)
$A^c = \{x \in X : \ x \notin A\}$ (complement of A)

Definition 1.1.2. The cartesian product $A \times B$ of two sets A, B, is the the set of all ordered pairs (a, b) with components $a \in A, b \in B$.
If $(x, y), (a, b)$ are in $A \times B$ then we have: $(x, y) = (a, b) \iff x = a$ and $y = b$.

Let us point out the following properties:

Proposition 1.1.3. Let A, B, C be subsets of X, then we have:

$$A \cup B = B \cup A$$
$$A \cap B = B \cap A$$
$$A \cup (B \cup C) = (A \cup B) \cup C$$
$$A \cap (B \cap C) = (A \cap B) \cap C$$
$$A \cap (B \cup C) = (A \cap B) \cup (A \cap C)$$
$$A \cup (B \cap C) = (A \cup B) \cap (A \cup C)$$
$$(A \cup B)^c = A^c \cap B^c$$
$$(A \cap B)^c = A^c \cup B^c$$

Proof: Straightforward

In the sequel we will extend these operations to arbitrary families of sets.

2. Graphs, Binary Relations, Functions

Definition 1.2.1. A graph between two sets X and Y is defined by a subset G of the catesian product $X \times Y$. Any graph G on $X \times Y$ defines a binary relation R between the two sets X and Y Any pair $(x, y) \in G$, reads y is $R-$related to x, and may be denoted by xRy.
In particular a binary relation on a set A is a subset of $A \times A$.
The domain of a graph G is the set $D = \{x \in X : \ \exists y \in Y, (x, y) \in G\}$. Similarly, the range of G is $C = \{y \in Y : \ \exists x \in X, (x, y) \in G\}$.

Definition 1.2.2. (a) We say that a graph G on $X \times Y$ is a functional graph if for each $x \in X$, there is at most one $y \in Y$ such that $(x, y) \in G$.

1

(b) Any functional graph G defines a function (or a mapping) f from X into Y according to the following recipe:

$$f : X \longrightarrow Y, \quad x \in X, y = f(x)$$

where $y \in Y$, is the unique element such that $(x, y) \in G$. Since we will deal, mainly with functions $f : X \longrightarrow Y$ rather than graphs, we agree to consider X as the domain, and to call the set Y the codomain of the function. The range of a function $f : X \longrightarrow Y$ is the range of its graph and can be written as

$$\text{Range}\,(f) = \{f(x) : x \in X\}.$$

(c) If $f : X \longrightarrow Y$ and $g : Y \longrightarrow Z$ are functions, the composition $g \circ f : X \longrightarrow Z$ is defined by $(g \circ f)(x) = g(f(x))$, $x \in X$.

Definition 1.2.3. Let $f : X \longrightarrow Y$ be a function

    (a) f is one-to-one (or injective) if it satisfies: $\forall a, b \in X$, $f(a) = f(b) \Longrightarrow a = b$

    (b) f is onto (or surjective) if $\forall y \in Y$, $\exists x \in X$, such that $y = f(x)$.

    (c) f is bijective if it is one-to-one and onto.

Proposition 1.2.4. Let $f : X \longrightarrow Y$ be a bijective function, then there exists a function $f^{-1} : Y \longrightarrow X$ such that:

$$\forall y \in Y, \left(f \circ f^{-1}\right)(y) = y \text{ and } \forall x \in X, \left(f^{-1} \circ f\right)(x) = x$$

f^{-1} is called the inverse function of the bijection f.

Proof. Let $y \in Y$, then there is $x \in X$ such that $y = f(x)$, because f is onto and such x is unique because f is one-to-one. Define $f^{-1} : Y \longrightarrow X$ by $f^{-1}(y) = x$. so we have $f^{-1}(y) = x \Longleftrightarrow .y = f(x)$. It is clear that f^{-1} is well defined and satisfies the property announced.∎

Definition 1.2.5. Let X be a set, and let $\mathcal{P}(X)$ be the power set of X. If I is any nonempty set, a function $f : I \longrightarrow \mathcal{P}(X)$ defines a family $\{A_i,\ i \in I\}$ of subsets of X, with $A_i = f(i) \in \mathcal{P}(X)$. For such family we perform the union and the intersection by:

$$\underset{i}{\cup}\, A_i = \{x : \exists i \in I,\ x \in A_i\}$$
$$\underset{i}{\cap}\, A_i = \{x : \forall i \in I,\ x \in A_i\}$$

Axiom of choice 1.2.6. Let $\{X_i,\ i \in I\}$ be a family of nonempty sets, then there is a function $\varphi : I \longrightarrow \underset{i}{\cup} X_i$ such that $\varphi(i) \in X_i$. The function φ permits to choose in each set X_i an element $x_i = \varphi(i)$.

If one is concerned with a finite family of sets $X_i, i = 1, 2 ... n$, such choice function is easily defined.

Definition 1.2.7. Let $\{X_i,\ i \in I\}$ be a family of sets and let F be the set of all functions, $f : I \longrightarrow \underset{i}{\cup} X_i$ such that $f(i) \in X_i$ for each $i \in I$. If none of the X_i is empty, the set F is not empty, by the Axiom of choice, and defines the direct product of the sets X_i. It will be denoted by $\underset{i}{\Pi} X_i$. So each $x \in \underset{i}{\Pi} X_i$ can be written as $x = (x_i)$, with $x_i = f(i) \in X_i$, $\forall i \in I$. If we have $X_i = X$, $\forall i \in I$, for some set X, the direct product is denoted by $\underset{i}{\Pi} X_i = X^I$.

As for the equality in $\underset{i}{\Pi} X_i$, it is defined by:

$$x = (x_i),\, y = (y_i) \in \underset{i}{\Pi} X_i, \quad x = y \Longleftrightarrow x_i = y_i,\, \forall i \in I.$$

3. Exercises

1. Let $f : X \longrightarrow Y$ be a function. Prove that:
(a) f is surjective iff there is a function $s : Y \longrightarrow X$ such that $f \circ s = I_Y$
    where $I_Y : Y \longrightarrow Y$ is the identity function of Y.
(b) f is injective iff there is a function $r : Y \longrightarrow X$ such that $r \circ f = I_X$
    where $I_X : X \longrightarrow X$ is the identity function of X.

2. Let $f : X \longrightarrow Y$ be a function. If $A \subset X$, and $B \subset Y$
put $f(A) = \{f(x),\, x \in A\}$, and $f^{-1}(B) = \{x \in X : f(x) \in B\}$. $f(A)$ is the direct
image of A, and $f^{-1}(B)$ is the inverse image of B.
Prove that:
(a) $A \subset f^{-1}(f(A))$, with equality if f is injective.
(b) $f(f^{-1}(B)) \subset B$, with equality if f is surjective.

3. Let $f : X \longrightarrow Y$, $g : Y \longrightarrow Z$ be bijective functions. Show that $g \circ f$ is bijective
and we have $(g \circ f)^{-1} = f^{-1} \circ g^{-1}$.

4. Let $f : X \longrightarrow Y$, be a function and $\{A_i,\, i \in I\}$ a family of subsets of X
$\{B_i,\, i \in I\}$ a family of subsets of Y, Prove the following properties:
(a) $f^{-1}\left(\underset{i}{\cup}B_i\right) = \underset{i}{\cup} f^{-1}(B_i), \quad f^{-1}\left(\underset{i}{\cap}B_i\right) = \underset{i}{\cap} f^{-1}(B_i)$
(b) $f\left(\underset{i}{\cup}A_i\right) = \underset{i}{\cup} f(A_i)$
(c) $f\left(\underset{i}{\cap}A_i\right) \subset \underset{i}{\cap} f(A_i)$, with equality if f is injective
(d) $\left(\underset{i}{\cup}A_i\right)^c = \underset{i}{\cap}A_i^c, \quad \left(\underset{i}{\cap}A_i\right)^c = \underset{i}{\cup}A_i^c, \quad$ (De Morgan's Law)

4. Binary Relations Properties

Definition 1.4.1. Let R be a binary relation on a set X, with graph G (see
definition **1.2.1**),and whose domain is X.
  R is reflexive if $\forall x \in X\ (x,x) \in G$
  R is symmetric if $\forall x, y \in X, (x,y) \in G \Longrightarrow (y,x) \in G$
  R is transitive if $\forall x, y, z \in X, (x,y) \in G$ and $(y,z) \in G \Longrightarrow (x,z) \in G$
  R is antisymmetric $\forall x \neq y \in X, (x,y) \in G \Longrightarrow (y,x) \notin G$

Definition 1.4.2. Let R be a binary relation on a set X, whose graph is G. We
say that R is an equivalence relation if it is reflexive, symmetric and transitive.
  For each $x \in X$, the equivalence class of x is the set
$$C_x = \{y \in X : (x,y) \in G\}$$
The quotient of the set X by the equivalence relation R is the set
$$X/R = \{C_x, x \in X\}.$$

Definition 1.4.3. If R is an equivalence relation on a set X, with quotient X/R,
we define the canonical mapping $p : X \longrightarrow X/R$, by $p(x) = C_x$, $x \in X$.

Proposition 1.4.4. Let R be an equivalence relation on a set X, then we have:
(a) For each $x, y \in X$, either $C_x = C_y$, or $C_x \cap C_y = \emptyset$
(b) $X = \underset{x}{\cup}C_x$, i.e the classes C_x form a partition of X
(c) The canonical mapping $p : X \longrightarrow X/R$, is onto
Proof. (a) If $C_x \cap C_y \neq \emptyset$, there is $z \in C_x \cap C_y$, then $(x,z) \in G$ and $(z,y) \in G$,
so, by transitivity $(x,y) \in G$. But in this case $C_x = C_y$

(b) is trivial

(c) comes from the definition of the canonical mapping p.∎

Definition 1.4.5. Let R be a binary relation on a set X. The relation R is called an ordering on X if it is reflexive, transitive and antisymmetric.

An ordering R on X is total if for each $x, y \in X$, either xRy or yRx is true, otherwise R, is said to be partial.

Example 1.4.6.

(a). The field $\mathbb{R}$ of the real numbers is totally ordered by the usual ordering:
$$x, y \in \mathbb{R},\ x \leq y \text{ (for the construction of the field } \mathbb{R}, \text{ see } [6]).$$
We will study this ordering in some details later.

(b). The power set $\mathcal{P}(X)$ of any set X is partially ordered by the inclusion:
$$A, B \in \mathcal{P}(X),\ A \subset B.$$

(c). The ring $\mathbb{Z}$ of the integers is partially ordered by the division R :
$$x, y \in \mathbb{Z},\ xRy \iff x \text{ divides } y.$$
In the sequel we make the convention to denote an ordered set by $X, \leq$

Definition 1.4.7. Let A be a subset of an ordered set $X, \leq$ and let $a \in X$.

a is an upper bound (resp. a lower bound) of A if $x \leq a$ (resp. $a \leq x$), $\forall x \in A$. We say that A is bounded above (resp. bounded below) if it has an upper bound (resp. a lower bound) A is bounded if it is bounded above and below.

Definition 1.4.8. Let A be a subset of an ordered set $X, \leq$ and let $a \in X$.

We say that a is a supremum of A if it is an upper bound and if for any upper bound b of A we have $a \leq b$. In other words a is a smallest upper bound. We denote a supremum by $a = \sup A$.

If $\sup A \in A$, we say that $\sup A$ is a maximal element of A.

We say that a is an infimum of A if it is a lower bound and if for any lower bound b of A we have $b \leq a$. In other words a is a greatest lower bound. We denote an infimum by $a = \inf A$.

If $\inf A \in A$, we say that $\inf A$ is a minimal element of A.

Definition 1.4.9. A chain in an ordered set $X, \leq$ is a totally ordered subset of X. A maximal chain in X is a chain which is not contained in any other chain of X.

Proposition 1.4.10. Zorn's lemma

Let $X, \leq$ be an ordered set in which every chain has an upper bound in X.

Then X has at least one maximal element

Proposition 1.4.11. Hausdorff Maximal Principle

Every ordered set $X, \leq$ contains a maximal chain.

The Hausdorff maximal principle states that, in any partially ordered set, every totally ordered subset is contained in a maximal totally ordered subset.

Theorem 1.4.12. The following principles are equivalent:

(a) Zorn's lemma

(b) Hausdorff Maximal Principle

(c) Axiom of choice

5. Exercises

5. Fix $m \in \mathbb{Z}$, $m \neq 0$, and consider the relation R on $\mathbb{Z}$ given by:
$p, q \in \mathbb{Z}$, $pRq \iff \exists k \in \mathbb{Z}$, $p = q + m.k$ (equality modulo m)
Prove that R is an equivalence relation and find the class C_q of any $q \in \mathbb{Z}$

6. A totally ordered set $X, \leq$ is said well ordered if every non empty subset $A \subset X$ has a minimal element.
(1) Prove that $\mathbb{N}$, equipped with its usual ordering, is well ordered
(2) The induction principle
Let $P(n), n \in \mathbb{N}$, $n \geq 1$, be a statement depending on n. Suppose that:
(a) $P(1)$ is true
(b) If $P(n)$ is true then $P(n+1)$ is true
Prove that $P(n)$ is true for every $n \geq 1$
Hint: Consider the set $E = \{n \geq 1 : P(n) \text{ is false}\}$, assume $E \neq \emptyset$ and use the fact that $\mathbb{N}$ is well ordered to get a contradiction.

7. Let G be a group and $H \subset G$ be a subgroup of G. Let R be the relation on G defined by:

$$x, y \in G, \ xRy \iff x.y^{-1} \in H, \ \left(y^{-1} \text{ is the inverse of } y\right)$$

Prove that:
(a) R is an equivalence relation on G
(b) For $x \in G$, the class of x is the set $Hx = \{h.x : h \in H\}$

6. The Real Number System

Theorem 1.6.1. There exists a totally ordered field $\mathbb{R}$, containing the set $\mathbb{Q}$ of rational numbers as a subfield and having the following properties:
    (1) Every subset $A \subset \mathbb{R}$ bounded above, has a supremum in $\mathbb{R}$
    (2) The equation $x^2 = 2$ has a solution in $\mathbb{R}$
    (see [6] for the proof)

Let us recall that $\mathbb{Q}$ does not possess these properties and $\mathbb{R}$ is in some sens the completion of $\mathbb{Q}$. Let us quote the following important facts about the field $\mathbb{R}$:

Proposition 1.6.2. (a) Archimedian Property: For any $x, y > 0$, there is an integer $n \geq 1$ such that $nx > y$.
    (b) The set $\mathbb{Q}$ of rational numbers is dense in $\mathbb{R}$.

Proposition 1.6.3. Let $A \subset \mathbb{R}$ be bounded above and let $\alpha = \sup A$, then:
    (a_1) α is an upper bound for A
    (a_2) $\forall \epsilon > 0, \exists x \in A : \alpha - \epsilon < x \leq \alpha$
the property (a_2) comes from the fact that α is the least upper bound for A.

Similarly if A is bounded below and if $\beta = \inf A$, then:
    (b_1) β is a lower bound for A
    (b_2) $\forall \epsilon > 0, \exists x \in A : \beta \leq x < \beta + \epsilon$
where (b_2) comes from the fact that β is the greatest lower bound for A.

Remark 1.6.4. Taking $\epsilon = \dfrac{1}{n}, n \geq 1$ in $(a_2) - (b_2)$, it is easy to construct two convergent sequences a_n, b_n in A such that $a_n \longrightarrow \alpha$ and $b_n \longrightarrow \beta$

7. Exercises

8. Let $f : X \longrightarrow \mathbb{R}$ be a function from a set $X \neq \emptyset$ into $\mathbb{R}$ such that $f(X)$ is bounded,with $a = \sup f(X)$ and $b = \inf f(X)$.

Prove that there are two sequences a_n, b_n in X such that $\lim_n f(a_n) = a$ and $\lim_n f(b_n) = b$.

9. Prove that For any $x, y \in \mathbb{R}$, there is an integer $n \in \mathbb{Z}$ such that $nx > y$.

Hint: use Proposition **1.6.3** (a) for $x < 0 < y$ and $x < y < 0$.

10. Let A, B be subsets of $\mathbb{R}$ and let us define the subsets:

$A + B = \{a + b : a \in A, b \in B\}$

$A - B = \{a - b : a \in A, b \in B\}$

$-A = \{-a : a \in A\}$

Assume that A and B are bounded. Prove that:

$\inf(-A) = -\sup A$, $\sup(-A) = -\inf A$

$\sup(A + B) = \sup A + \sup B$, $\sup(A - B) = \sup A - \inf B$

$\inf(A + B) = \inf A + \inf B$, $\inf(A - B) = \inf A - \sup B$

11. Consider the set $\mathbb{R}\backslash\mathbb{Q}$ of irrational numbers. Prove that:

(a) $r \in \mathbb{Q}$ and $v \in \mathbb{R}\backslash\mathbb{Q} \Longrightarrow r.v \in \mathbb{R}\backslash\mathbb{Q}$ and $\dfrac{1}{v} \in \mathbb{R}\backslash\mathbb{Q}$

(b) $\forall x, y \in \mathbb{R}, x < y \Longrightarrow \exists w \in \mathbb{R}\backslash\mathbb{Q} : x < w < y$

$((b)$ means that $\mathbb{R}\backslash\mathbb{Q}$ is dense in $\mathbb{R})$

[take $v \in \mathbb{R}, v > 0$, so $vx < vy$; if $r \in \mathbb{Q}$ and $vx < r < vy$, the number $w = \dfrac{r}{v}$ is in $\mathbb{R}\backslash\mathbb{Q}$ by (a) and works]

8. Cardinals

Let X be a set with power set $\mathcal{P}(X)$

Definition 1.8.1. The sets $E, F \in \mathcal{P}(X)$ are called equipotent if there is a bijection $f : E \longrightarrow F$ from E onto F.We denote this relation by $E \approx F$.

It is easy to check that the binary relation $\approx$ is an equivalence relation on $\mathcal{P}(X)$. The equivalence class of E is called the Cardinal of E and denoted by $\mathrm{Card}(E)$.

Definition 1.8.2. Let n be a positive integer and let E_n be a set with n elements:

(a) A set E is finite if there is n such that $E \approx E_n$ and we have

$E_m \approx E_n \iff m = n$.

(b) A set E is infinite if for each finite subset $F \subset E$ the set $E\backslash F$ is not empty.

(c) A set E is infinite countable if there is a bijection $f : \mathbb{N} \longrightarrow E$

from $\mathbb{N}$ onto E. If $f(n) = x_n$, we have $E = \{x_1, x_2, ..., x_n, ...\}$.

(d) A set E is uncountable if for each countable subset $F \subset E$ the set $E\backslash F$ is not empty.

(e) A set E is at most countable if it is finite or infinite countable.

Example: The set $\mathbb{Z}$ of integers is countable. Indeed, define the bijection $f : \mathbb{N} \longrightarrow \mathbb{Z}$ by:

$f(2n + 1) = n$ and $f(2n) = -n$

Proposition 1.8.3. Let E be a countable infinite set and let $A \subset E$ be an infinite subset of E. Then A is countable.

Proof: Put $E = \{x_1, x_2, ..., x_n, ...\}$, and define the sequence

$n_1, n_2, ...$ of integers by:

$n_1 = \inf\{n \geq 1, x_n \in A\}$

$n_2 = \inf\{n > n_1, x_n \in A\}$

$n_3 = \inf\{n > n_2, x_n \in A\}$ ....and more generally $n_k = \inf\{n > n_{k-1}, x_n \in A\}$.

Then we have $1 \leq n_1 < n_2 < n_3 < ... < n_k < ...$,and $A = \{x_{n_1}, x_{n_2}, ..., x_{n_k}, ...\}$.

The function $f : \mathbb{N} \longrightarrow A$, given by $f(k) = x_{n_k}$, $k \geq 1$ is a bijection, as may be seen from the construction of the sequence (n_k). So A is countable.■

Corollary 1 Every subset $A \subset \mathbb{N}$ is at most countable.

Corollary 2 Let E be a set. If there is an injective function $f : E \longrightarrow \mathbb{N}$, then E is countable.

Indeed $f(E) \subset \mathbb{N}$ is countable and equipotent to E.

Proposition 1.8.4. Let E be a set. If there is an surjective function $f : \mathbb{N} \longrightarrow E$, then E is countable.

Proof: Let $x \in E$, then $A_x = f^{-1}(x) \subset \mathbb{N}$ is not empty and since $\mathbb{N}$ is well ordered (see exercise **6**), $\inf A_x = \varphi(x)$ exists and $\varphi(x) \in f^{-1}(x)$. So we have $f \circ \varphi(x) = x$, and then the function φ is injective from E into $\mathbb{N}$(see exercise1), by corollary **2** above E is countable.■

Theorem 1.8.5. The Cartesian product $\mathbb{N} \times \mathbb{N}$ is countable.

Proof: Define the function $f : \mathbb{N} \longrightarrow \mathbb{N} \times \mathbb{N}$ by giving $f(1) = (1,1)$ and

if $f(n) = (1,k)$ then $f(n+1) = (k+1,1)$

if $f(n) = (r,k)$,with $r \neq 1$ then $f(n+1) = (r-1,k+1)$

$(f(2) = (2,1), f(3) = (1,2), f(4) = (3,1), f(5) = (2,2)....)$

it is not difficult to check that the function so defined is bijective.■

Corollary If E_1, E_2 are infinite countable sets, then the cartesian product $E_1 \times E_2$ is countable. More generally if $E_1, E_2, ..., E_n$ is a finite family of infinite countable sets, then the product $E_1 \times E_2 \times ... \times E_n$ is countable.

Proof: Let $f : E_1 \longrightarrow \mathbb{N}$, $g : E_2 \longrightarrow \mathbb{N}$, $h : \mathbb{N} \times \mathbb{N} \longrightarrow \mathbb{N}$ be bijective functions and let us define $\varphi : E_1 \times E_2 \longrightarrow \mathbb{N} \times \mathbb{N}$ by $\varphi(a,b) = (f(a), g(b))$. It is clear that the functions φ and $h \circ \varphi$ are bijective, this yields that $E_1 \times E_2$ is countable. For n sets use induction method (exercise **6**).■

This corollary is not true for infinite product of sets, as is shown by the following example:

Example 1.8.6: Let E be the infinite product set $E = \{0,1\}^{\mathbb{N}}$. We prove that E is uncountable. It is enough to prove that for any infinite countable set $A \subset E$, the set $E \backslash A$ is nonempty. Let $A = \{a_1, a_2, ..., a_n, ...\}$ be such countable set. Since $a_n \in E$, put $a_n = (a_{n1}, a_{n2}, ..., a_{nm}, ...)$, where $a_{nm} = 0$ or 1. Now define b in $\{0,1\}^{\mathbb{N}}$ by $b = (y_1, y_2, ..., y_n, ...)$, where $y_n = 1 - a_{nn}$. Then it is clear that $b \neq a_n$ for every n, so $b \in E \backslash A$.

Theorem 1.8.7. Let E_n, $n \in \mathbb{N}$ be a sequence of countable sets, then the union $E = \underset{n}{\cup} E_n$, is countable.

Proof: By Proposition **1.8.4**, it is enough to construct a surjective function from $\mathbb{N}$ onto E. Let $f : \mathbb{N} \longrightarrow \mathbb{N} \times \mathbb{N}$ be the bijection of Theorem **1.8.5**, and let f_j be a bijection from $\mathbb{N}$ onto E. Define $\varphi : \mathbb{N} \times \mathbb{N} \longrightarrow E$, by $\varphi(i,j) = f_j(i)$. The function φ is surjective, indeed if $x \in E$ there is $j \in \mathbb{N}$ such that $x \in E_j$ and, since f_j is a bijection, there is $i \in \mathbb{N}$ with $x = f_j(i)$. Now the function $\varphi \circ f : \mathbb{N} \longrightarrow E$ is surjective, so E is countable. $\blacksquare$

9. Exercises

12. Prove that the power set $\mathcal{P}(\mathbb{N})$ of $\mathbb{N}$ is uncountable.
[consider the function $f : \mathcal{P}(\mathbb{N}) \longrightarrow \{0,1\}^{\mathbb{N}}$, given by $f(A) = (x_n)$
with $x_n = 1 \iff n \in A$, show that f is a bijection and use example **1.8.6**]

13. Let $\mathcal{P}_0(\mathbb{N})$ be the family of finite subsets of $\mathbb{N}$.
Prove that $\mathcal{P}_0(\mathbb{N})$ is countable.

CHAPTER 2

Topological Structures

1. Topological Spaces

Definition 2.1.1. Let X be a set and let τ be a family of subsets of X. We say that τ defines a topology on X if the following conditions are satisfied:

(1) The subsets X and $\emptyset$ are in τ

(2) The union of every subfamily of sets in τ, is in τ

(3) The intersection of every finite subfamily of sets in τ, is in τ

The sets in such family τ are called open sets. The set X equipped with τ is called a topological space.

Examples 2.1.2. (a) For any set X put $\tau = \{X, \emptyset\}$. This is a topology with $X, \emptyset$ as the only open sets. It is called the trivial topology and denoted by τ_0.

(b) The family $\tau = \mathcal{P}(X)$ of all subsets of X defines a topology called the discrete topology and denoted by τ_1. Every singleton $\{x\}$ is open for τ_1.

(c) Let $X = \mathbb{R}$ and define τ by: $G \in \tau \iff G = \underset{\alpha \in A}{\cup} I_\alpha$ where $\{I_\alpha, \alpha \in A\}$ is a family of open intervals. Then τ satisfies the conditions of Definition **2.1.1.** Indeed, $(1), (2)$ are immediate, to see (3), let $G_1, G_2, ..., G_n$, be a finite family in τ, with $G_i = \underset{\alpha_i \in A_i}{\cup} I_{\alpha_i}$, $i = 1, 2, ..., n$.

So we have $G_1 \cap G_2 \cap ... \cap G_n = \underset{(\alpha_1,..,\alpha_n) \in A_1 \times .. \times A_n}{\cup} I_{\alpha_1} \cap .. \cap I_{\alpha_n}$, and $I_{\alpha_1} \cap .. \cap I_{\alpha_n}$

is either empty or an open interval.

(d) In the euclidean space $X = \mathbb{R}^n$, we define an open rectangle as a set of the form $R = I_1 \times I_2 \times .. \times I_n$, where I_k ,$k = 1, 2, ..., n$, are open intervals of $\mathbb{R}$. Define τ as follows: $G \in \tau \iff G = \underset{\alpha \in A}{\cup} R_\alpha$ where $\{R_\alpha, \alpha \in A\}$ is a family of open rectangles. It is not difficult to check that such τ defines a topology on $\mathbb{R}^n$, called the euclidean topology.

Definition 2.1.3. Let X, τ be a topological space and let $\mathcal{B} \subset \tau$ be a subfamily of open sets. We say that $\mathcal{B}$ is a base for τ if every open set in τ can be written as a union of open sets in $\mathcal{B}$.

Examples 2.1.4. (a) The family of open intervals (resp. the family of open rectangles) is a base for the topology of $\mathbb{R}$ in **2.1.2.**(c) (resp. for the euclidean topology of $\mathbb{R}^n$).

(b) In every topological space X, τ, the family τ is a base for τ.

Proposition 2.1.5. Let X, τ be a topological space and let $\mathcal{B} \subset \tau$ be a subfamily of open sets. Then $\mathcal{B}$ is a base iff it satisfies the following condition:

For every $G \in \tau$ and every $x \in G$, there is $B \in \mathcal{B}$ such that $x \in B \subset G$.

9

Proof: If $\mathcal{B}$ is a base and if $G \in \tau$, there is a family $\{B_i, i \in I\} \subset \mathcal{B}$ such that $G = \bigcup_i B_i$; so if $x \in G$, $\exists i$ with $x \in B_i \subset G$. To see the converse let $G \in \tau$, then, by the condition, for each $x \in G$, there is $B_x \in \mathcal{B}$ such that $x \in B_x \subset G$. But in this case we have $G = \bigcup_{x \in G} B_x$, consequently $\mathcal{B}$ is a base.∎

Definition 2.1.6. Let X, τ be a topological space and let $A \subset X$. Let $x \in A$, we say that A is a neighborhood of x if there is an open set $G \in \tau$ such that $x \in G \subset A$. If A is a neighborhood of x, we call x an interior point of A. We denote the set of all interior points of A by $\overset{o}{A}$ and call it the interior of A.

Proposition 2.1.7. Let X, τ be a topological space and let $A \subset X$. then:
    (a) A is open $\iff$ A is a neighborhood of each of its points.

    (b) The interior set $\overset{o}{A}$ is open for each A.

Proof: (a) The part $\implies$ is trivial. Conversely, if A is a neighborhood of each of its points, for each $x \in A$ there is an open set G_x such that $x \in G_x \subset A$. But in this case $A = \bigcup_{x \in A} G_x$, and any union of open sets is open, so A is open.

(b) Let $x \in \overset{o}{A}$, there is $G \in \tau$ such that $x \in G \subset A$, so each $y \in G$ is in $\overset{o}{A}$ and then $G \subset \overset{o}{A}$, consequently $\overset{o}{A}$ is open by Proposition **2.1.5** and the fact that τ is a base for τ.∎

Since any neighborhood contains an open neighborhood, we will consider in the sequel only open neighborhoods.

Definition 2.1.8. A family $\mathcal{V}(x)$ of neighborhoods of a point x in a space X, τ, is said to be a base at x if for every neighborhood V of x there is $U \in \mathcal{V}(x)$ such that $U \subset V$.

As examples one can cite:

1. In a discrete space X, τ_1, the family $\mathcal{V}(x)$ of finite subsets containing x is a base at x.

2. In $\mathbb{R}^n$ endowed with the euclidean topology, the family of open rectangles containing x is a base at x.

Proposition 2.1.9. Let $\mathcal{B}$ be a family of open sets in a space X, τ.
Then $\mathcal{B}$ is a base for τ iff $\mathcal{B}$ satisfies the following condition:
    For each $x \in X$ the family $\mathcal{B}(x) = \{B \in \mathcal{B} : x \in B\}$ is a base at x.

Proof: Suppose that $\mathcal{B}$ is a base for τ and let $x \in X$. If V is a neighborhood of x, there is an open set U with $x \in U \subset V$. By Proposition **2.1.5.**, there is $B \in \mathcal{B}$ such that $x \in B \subset U$ and then $x \in B \subset V$, so $B \in \mathcal{B}(x)$ and $\mathcal{B}(x)$ is a base at x. Conversely, suppose that $\mathcal{B}(x)$ is a base at x for each x and let G be an open set, then for each $x \in G$, there $B_x \in \mathcal{B}(x)$ with $x \in B_x \subset G$, because G is a neighborhood of x. So we deduce that $G = \bigcup_{x \in G} B_x$
consequently $\mathcal{B}$ is a base for τ.∎

Definition 2.1.10. (Comparison of topologies)
    Let τ and σ be two topologies on the same set X. We say that τ is thinner than σ (or that σ is coaser than τ) if $\tau \subset \sigma$.
We can easily check that this binary relation is a partial ordering on the family $\mathcal{T}(X)$ of all topologies on X, denoted by $\tau \prec \sigma$.

Examples 2.1.11. (a) Let $\{\tau_\alpha : \alpha \in A\}$ be a family of topologies on X, and put $\tau = \underset{\alpha \in A}{\cap} \tau_\alpha$, then we have $\tau \prec \tau_\alpha$ for all α. In other words τ is a lower bound for the family $\{\tau_\alpha : \alpha \in A\}$. We will see that the family has also an upper bound.
(b) Let τ_0 be the trivial topology and τ_1 the discrete topology on X, then for every $\tau \in \mathcal{T}(X)$ we have $\tau_0 \prec \tau \prec \tau_1$, in other words τ_0 is minimal and τ_1 is maximal in $\mathcal{T}(X)$.

Theorem 2.1.12. (Topology generated by a family of sets)
Let $\mathcal{F}$ be a family of subsets of a set X. Then, among all topologies on X containing $\mathcal{F}$ there is a smallest one, called the topology generated by $\mathcal{F}$ and denoted by $\tau(\mathcal{F})$. This topology can be described explicitly as follows:
Let $\mathfrak{I}$ be the family formed by $X, \emptyset$ and all finite intersections of elements in $\mathcal{F}$, then $\tau(\mathcal{F})$ is exactly the family of all unions of elements of $\mathfrak{I}$. In other words $\mathfrak{I}$ is a base for $\tau(\mathcal{F})$.

Proof: Let $\varsigma = \{\sigma \in \mathcal{T}(X) : \mathcal{F} \subset \sigma\}$ then ς is not empty since $\tau_1 = \mathcal{P}(X) \in \varsigma$. By example **2.1.11.**(a), we have $\tau(\mathcal{F}) = \underset{\sigma \in \varsigma}{\cap} \sigma$. To prove that $\tau(\mathcal{F})$ has the descripton given in the theorem let us point out that the family $\widetilde{\mathfrak{I}}$ of all unions of elements of $\mathfrak{I}$ is itself a topology, as may be easily checked. Note that $\mathfrak{I} \subset \tau(\mathcal{F})$ and so $\widetilde{\mathfrak{I}} \subset \tau(\mathcal{F})$. Finally from the trivial inclusion $\mathcal{F} \subset \mathfrak{I}$ we deduce that $\tau(\mathcal{F}) \subset \widetilde{\mathfrak{I}}$ and then $\tau(\mathcal{F}) = \widetilde{\mathfrak{I}}$ as wanted.∎

Definition 2.1.13. A family $\mathcal{F}$ of subsets of X is a subbase for a topology τ on X if $\tau = \tau(\mathcal{F})$
The following theorem gives a condition on $\mathcal{F}$ for being a base of $\tau(\mathcal{F})$.

Theorem 2.1.14. Suppose that the family $\mathcal{F}$ satisfies the following condition:
If F_1, F_2 are in $\mathcal{F}$ and if $x \in F_1 \cap F_2$, there is $F \in \mathcal{F}$ such that $x \in F \subset F_1 \cap F_2$
Then $\mathcal{F}$ is a base of $\tau(\mathcal{F})$.

Proof: With the notations of Theorem **2.1.12**, it is enough to prove that every $B \in \mathfrak{I}$ can be written as the union of elements of $\mathcal{F}$. By Proposition **2.1.5**, it is enough that for each $x \in B$ there is $F \in \mathcal{F}$ such that $x \in F \subset B$. But if $B \in \mathfrak{I}$ there exist $F_1, F_2, ..., F_n \in \mathcal{F}$, such that $B = F_1 \cap F_2 \cap ... \cap F_n$. For $n = 2$ the condition of the Theorem provides $F \in \mathcal{F}$ such that $x \in F \subset B$. For arbitrary n use induction.∎

Examples 2.1.15. (a) Let $(X, \tau), (Y, \sigma)$ be two topological spaces, and consider on the product $Z = X \times Y$ the class of subsets $\mathcal{F} = \{A \times Y, X \times B : A \in \tau, B \in \sigma\}$. Then the family $\mathfrak{I}$ of Theorem **2.1.12** is given by $\mathfrak{I} = \{A \times B : A \in \tau, B \in \sigma\}$. The topology $\tau(\mathcal{F})$ on Z is called the product topology of τ and σ and is denoted by $\tau \otimes \sigma$. The family $\mathfrak{I}$ is a base for $\tau \otimes \sigma$.

(b) Let $X = C[0,1]$ be the set of all continuous functions $f : [0,1] \longrightarrow \mathbb{R}$. For each $f \in X$ and each $\epsilon > 0$, put $V(f, \epsilon) = \left\{ g \in X : \int_{[0,1]} |f - g| < \epsilon \right\}$, where the integral is a Riemann one. Now consider the family: $\mathcal{V} = \{V(f, \epsilon) : f \in X, \epsilon > 0\}$. Using Theorem **2.1.14**, we show that V is a base for a topology on X. We have to check the condition of the Theorem for the family $\mathcal{V}$. Let $f, g \in X$, and $\epsilon, \epsilon' > 0$; if $h \in V(f, \epsilon) \cap V\left(g, \epsilon'\right)$, we prove the existence of $\epsilon'' > 0$ such that

$V\left(h,\epsilon''\right) \subset V\left(f,\epsilon\right) \cap V\left(g,\epsilon'\right)$. In order to do it, put $\alpha = \displaystyle\int_{[0,1]} |f - h|$ and

$\beta = \displaystyle\int_{[0,1]} |g - h|$, then taking $\epsilon'' = Min\left(\epsilon - \alpha, \epsilon' - \beta\right)$, we get

$V\left(h,\epsilon''\right) \subset V\left(f,\epsilon\right) \cap V\left(g,\epsilon'\right)$; indeed if $\varphi \in V\left(h,\epsilon''\right)$ then $\displaystyle\int_{[0,1]} |f - \varphi| \leq \displaystyle\int_{[0,1]} |f - h| +$

$\displaystyle\int_{[0,1]} |h - \varphi| < \alpha + \epsilon - \alpha = \epsilon$, so $\varphi \in V\left(f,\epsilon\right)$. Similarly, $\displaystyle\int_{[0,1]} |g - \varphi| \leq \displaystyle\int_{[0,1]} |g - h| +$

$\displaystyle\int_{[0,1]} |h - \varphi| < \beta + \epsilon' - \beta = \epsilon'$, whence $\varphi \in V\left(g,\epsilon'\right)$

(c) Let $\{\tau_\alpha : \alpha \in A\}$ be a family of topologies on a set X. We have seen in Example **2.1.11**(a) that $\tau = \bigcap_{\alpha \in A} \tau_\alpha$, is an infimum for the given family. We show that it has also a supremum denoted by $\bigvee_\alpha \tau_\alpha$. In fact $\bigvee_\alpha \tau_\alpha$ is the topology generated by the union $\bigcup_\alpha \tau_\alpha$. It is the smallest topology σ such that $\tau_\alpha \prec \sigma, \forall \alpha$, see **2.1.10** and **2.1.12**.

Definition 2.1.16. (Topological subspace)

Let (X, τ) be a topological space and let $A \subset X$ be a subset of X. Let us define the family τ_A of subsets of A by $\tau_A = \{G \cap A, G \in \tau\}$. Then it is easy to check that τ_A satisfies the axioms (1)-(2)-(3) of Definition **2.1.1** and so it defines a topology on A, called trace topology of τ on A. The set A endowed with τ_A is called topological subspace of X, τ. Any open set U of τ_A is of the form $U = G \cap A$, for some $G \in \tau$, that is the trace of an open set $G \in \tau$ on A. Such U is not open in X, however we have:

Proposition 2.1.17. The following statements are equivalent:

    (a) Any open set in A, τ_A, is open in X, τ

    (b) The subset A is open in X, τ

The proof is left to the reader.

Definition 2.1.18. Closed sets

    A subset F of a topological space X, τ is closed if the complement F^c is open

Examples 2.1.19. (a) In the discrete space (X, τ_1) every subset is closed.

(b) The half line $(-\infty, a]$ is closed in $\mathbb{R}$ endowed with the euclidean topology.

(c) The interval $(0, 1]$ is not closed (Why?)

(d) If $X = \{a, b\}$ and if $\tau = \{\emptyset, \{a\}, X\}$, then $\{a\}$ is not closed since $\{a\}^c = \{b\} \notin \tau$

Proposition 2.1.20. In a topological space X, τ, the closed sets satisfy the following properties:

    $(1)'$ X and $\emptyset$ are closed

    $(2)'$ The intersection of any family of closed sets is closed

    $(3)'$ The union of any finite family of closed sets is closed

Proof: These are the dual properties of the axioms (1)-(2)-(3) of Definition **2.1.1**.∎

Definition 2.1.21. The closure of a set A in X, τ is the smallest closed set in X containing A. It will be denoted by $\overline{A}$. To justify this definition, consider the family $\mathfrak{I} = \{F : F \text{ closed}, A \subset F\}$, then $\mathfrak{I}$ is not empty since $X \in \mathfrak{I}$ and we have $\overline{A} = \underset{F \in \mathfrak{I}}{\cap} F$. Let us point out that a set A is closed iff $A = \overline{A}$.

Theorem 2.1.22. Let A be a subset of X, τ and let $x \in X$, then the following statements are equivalent:

(a) $x \in \overline{A}$.

(b) $V \cap A \neq \emptyset$, for every open neighbohood V of x.

Proof: If (a) is not satisfied, we have $x \in V = \cup G$, the union being over all open $G \subset A^c$, but such V is an open neighbohood of x with $V \subset A^c$, so (b) is not satisfied. Conversely, if (b) is not satisfied, there is an open neighbohood V of x with $V \cap A = \emptyset$, this yields $A \subset V^c$ and $\overline{A} \subset V^c$, because V^c is closed, so $x \notin \overline{A}$, and (a) is not satisfied.∎

Definition 2.1.23. (a) Let A be a subset of X, τ. A point $x \in X$ is a limit point (or a cluster point) of A if for every open neighbohood V of x, $V \cap (A \backslash x) \neq \emptyset$. We denote by A' the set of all limit points of A.

(b) A point $x \in A$ is an isolated point if there is an open neighbohood V of x such that $V \cap A = \{x\}$.

Examples 2.1.24. Let $\mathbb{R}$ endowed with the euclidean topology and consider the subset $A = \left\{ \dfrac{1}{n}, n \geq 1 \right\}$. It is not difficult to check that every point of A is isolated and that $A' = \{0\}$.

Theorem 2.1.25. For every subset A of X, τ, we have $\overline{A} = A \cup A'$.

Proof: We have $A \subset \overline{A}$ and $A' \subset \overline{A}$, so $A \cup A' \subset \overline{A}$. On the otherhand if $x \in \overline{A}$, then either $x \in A$ or $x \notin A$. In the case $x \notin A$, we have $x \in A'$ since for every open neighborhood V of x, $V \cap (A \backslash x) \neq \emptyset$, by Theorem **2.1.22.** So in either case $x \in \overline{A} \implies x \in A \cup A'$∎

2. Exercises

14. Let X be an infinite set. Prove that the family $\tau = \{\emptyset; \; A \subset X : A^c \text{ finite}\}$ defines a topology on X.

15. Let X be a partially ordered set by the relation $\prec$. We consider the family τ of subsets G satisfying the condition:

If $x \in G$ then every $y \prec x$ is in G

Prove that τ defines a topology on X.

16. Let $X = C[0,1]$ be the set of all continuous functions $f : [0,1] \longrightarrow \mathbb{R}$. For each $f \in X$ and each $\epsilon > 0$, put $V(f, \epsilon) = \left\{ g \in X : \int_{[0,1]} |f - g| < \epsilon \right\}$, where the integral is a Riemann one. Now consider the family: $\mathcal{V} = \{V(f, \epsilon) : \; f \in X, \epsilon > 0\}$, and the family $\mathcal{U} = \{U(f, \epsilon) : \; f \in X, \epsilon > 0\}$, where $U(f, \epsilon) = \left\{ g : \; \underset{x}{Sup} |f(x) - g(x)| < \epsilon \right\}$.

Prove that $\mathcal{V}$ and $\mathcal{U}$ are bases of two topologies on $C\,[0,1]$. The case of $\mathcal{V}$ is considered in Example **2.1.15**(b). Prove that $\tau\,(\mathcal{V}) < \tau\,(\mathcal{U})$.

17. Prove that for any subsets A, B of a topological space X, τ, we have:

(a) $X\backslash\overline{A} = \overset{o}{\widehat{X\backslash A}}$

(b) $\overline{A \cup B} = \overline{A} \cup \overline{B}$

(c) $\overline{A \cap B} \subset \overline{A} \cap \overline{B}$

$\left(b'\right)$ $\overset{o}{\widehat{A \cap B}} = \overset{o}{A} \cap \overset{o}{B}$

$\left(c'\right)$ $\overset{o}{A} \cup \overset{o}{B} \subset \overset{o}{\widehat{A \cup B}}$

Prove that for any family $\{A_\alpha : A_\alpha \subset X,\ \alpha \in I\}$ we have:

(d) $\overline{\cap A_\alpha} \subset \cap \overline{A_\alpha}$
$\quad\;\,^\alpha \qquad\quad ^\alpha$

$\left(d'\right)$ If $\cup \overline{A_\alpha}$ is closed then $\overline{\cup A_\alpha} = \cup \overline{A_\alpha}$
$\qquad\quad\; ^\alpha \qquad\qquad\qquad\quad\; ^\alpha \qquad\; ^\alpha$

18. We define the boundary ∂A of a subset A in a topological space X, τ, by:

$$\partial A = \overline{A} \cap \overline{X\backslash A}$$

Let us observe that if $x \in \partial A$, then for every neighbohood V of x we have $V \cap A \neq \emptyset$ and $V \cap X\backslash A \neq \emptyset$.

(a) Prove that $\partial A = \emptyset$ iff A is open and closed.

(b) Show that if $\cap \partial B = \emptyset$, then $\overset{o}{\widehat{A \cup B}} = \overset{o}{A} \cup \overset{o}{B}$

(c) $\overline{A} = A \cup \partial A$

19. (a) Prove that the subset A is dense in X iff $\overset{o}{\widehat{X\backslash A}} = \emptyset$

$\quad\;\,(b)$ If A, B are open sets dense in X then $A \cap B$ is dense in X.

20. Let (X, τ), (Y, σ) be two topological spaces, and consider on the product $X \times Y$ the topology $\tau \otimes \sigma$ (see Example **2.1.15**). Prove that for any $A \in \tau$ and $B \in \sigma$ we have:

$$\overline{A \times B} = \overline{A} \times \overline{B}$$
$$\overset{o}{\widehat{A \times B}} = \overset{o}{A} \times \overset{o}{B}$$
$$\partial\,(A \times B) = \left(\partial A \times \overline{B}\right) \cup \left(\overline{A} \times \partial B\right)$$

3. Continuous Functions

Let $f : X \longrightarrow Y$ be a function. If $B \subset Y$ the inverse image of B is defined by $f^{-1}\,(B) = \{x \in X : f\,(x) \in B\}$ (see Exercise **1.3.2**)

Proposition 2.3.1. Let $f : X \longrightarrow Y$ be a function, then we have:
(a) For every topology σ on Y, the family $f^{-1}\,(\sigma) = \left\{f^{-1}\,(B) : B \in \sigma\right\}$
$\quad\;$ is a topology on X.
(b) For every topology τ on X, the family $\left\{B \subset Y : f^{-1}\,(B) \in \tau\right\}$
$\quad\;$ is a topology on Y.

Proof: It is a consequence of the inverse image properties given in Exercise **1.3.4**∎

Definition 2.3.2. Let (X, τ), (Y, σ) be two topological spaces and let $f : X \longrightarrow Y$ be a function.
(a) The function is said to be continuous if $f^{-1}\,(\sigma) \prec \tau$, that is
$\quad\; f^{-1}\,(G) \in \tau$, for every $G \in \sigma$.
(b) The function is said to be continuous at the point $x \in X$

if for every open neighbohood $V\left(f\left(x\right)\right)$ of $f\left(x\right)$ there is an
open neighbohood $U\left(x\right)$ of x such that $f\left(U\left(x\right)\right) \subset V\left(f\left(x\right)\right).$

Examples 2.3.3. (a) If $f : X \longrightarrow Y$ is the constant function, we have $f^{-1}\left(\sigma\right) = \{X, \emptyset\} \subset \tau$, so f is continuous.

(b) If A is a subset of X, τ, endowed with the trace topology τ_A then the canonical injection $i : A \longrightarrow X$ is continuous since we have
$$i^{-1}\left(G\right) = G \cap A \in \tau_A, \text{ for every } G \in \tau.$$

(c) Let X, τ be a topological space. If R is an equivalence relation on a set X, with quotient X/R, and if $p : X \longrightarrow X/R$ is the canonical mapping
(see Definition **1.4.2**), the family $\sigma_R = \{B \subset X/R : p^{-1}\left(B\right) \in \tau\}$ is a topology on X/R (Proposition **2.3.1.**(b)), and p is continuous from X, τ into $X/R, \sigma_R$.

(d) A function $f : X \longrightarrow Y$ from $\left(X, \tau\right)$ into $\left(Y, \sigma\right)$ is continuous iff the inverse image $f^{-1}\left(F\right)$ of any closed set F of Y is closed in X. This comes from the relation $f^{-1}\left(Y \backslash A\right) = X \backslash f^{-1}\left(A\right)$, valid for every subset $A \subset Y$.

Theorem 2.3.4. A function $f : X \longrightarrow Y$ from $\left(X, \tau\right)$ into $\left(Y, \sigma\right)$ is continuous iff it is continuous at each point x of X.

Proof: Suppose f continuous from $\left(X, \tau\right)$ into $\left(Y, \sigma\right)$, and let $x \in X$. Let $V\left(f\left(x\right)\right)$ be an open neighbohood of $f\left(x\right)$, then $U\left(x\right) = f^{-1}\left(V\left(f\left(x\right)\right)\right)$ is open in X, containing x and satisfies $f\left(U\left(x\right)\right) = f\left(f^{-1}\left(V\left(f\left(x\right)\right)\right)\right) \subset V\left(f\left(x\right)\right)$. Consequently f is continuous at x. Conversely suppose f continuous at each point x of X and let G be open in Y with $f^{-1}\left(G\right) \neq \emptyset$. Then G is an open neighborhood of $f\left(x\right)$, so there is an open neighborhood $U\left(x\right)$ of x such that $f\left(U\left(x\right)\right) \subset G$. We deduce that $U\left(x\right) \subset f^{-1}\left(f\left(U\left(x\right)\right)\right) \subset f^{-1}\left(G\right)$ and $i^{-1}\left(G\right)$ is a neighborhood of each of its points, by Proposition **2.1.7**(a), $f^{-1}\left(G\right)$ is open.∎

Proposition 2.3.5. Let $f : X \longrightarrow Y$ be continuous from $\left(X, \tau\right)$ into $\left(Y, \sigma\right)$ and $g : Y \longrightarrow Z$ continuous from $\left(Y, \sigma\right)$ into $\left(Z, \rho\right)$, then the composition $g \circ f : X \longrightarrow Z$ is continuous from $\left(X, \tau\right)$ into $\left(Z, \rho\right)$.

Proof: We have $\left(g \circ f\right)^{-1}\left(\rho\right) = f^{-1}\left(g^{-1}\left(\rho\right)\right) \subset f^{-1}\left(\sigma\right) \subset \tau$
where the first inclusion comes from the continuity of g and the second from the continuity of f.∎

Proposition 2.3.6. Let $f, g : X \longrightarrow \mathbb{R}$ be continuous functions from $\left(X, \tau\right)$ into $\mathbb{R}$ endowed with the euclidean topology. Then the following functions are continuous:
$$f + g : X \longrightarrow \mathbb{R} \ , \ \left(f + g\right)\left(x\right) = f\left(x\right) + g\left(x\right)$$
$$f.g : X \longrightarrow \mathbb{R} \ , \ \left(f.g\right)\left(x\right) = f\left(x\right).g\left(x\right)$$

For the proof we need the following lemma:
Lemma: A function is continuous iff for every $\alpha \in \mathbb{R}$ the sets $f^{-1}\,]\alpha, \infty[$ and $f^{-1}\,]-\infty, \alpha[$ are open in X.

Comes from the fact $f^{-1}\,]\alpha, \beta[= f^{-1}\,]\alpha, \infty[\cap f^{-1}\,]-\infty, \beta[$ and that $]\alpha, \infty[,]-\infty, \beta[$, are open.

Proof of 2.3.6: We have $\left(f + g\right)^{-1}\left(]\alpha, \infty[\right) = \underset{\beta \in \mathbb{R}}{\cup} \left(f^{-1}\left(]\alpha - \beta, \infty[\right) \cap g^{-1}\left(]\beta, \infty[\right)\right)$

and $\left(f + g\right)^{-1}\left(]-\infty, \alpha[\right) = \underset{\beta \in \mathbb{R}}{\cup} \left(f^{-1}\left(]-\infty, \alpha - \beta[\right) \cap g^{-1}\left(]-\infty, \beta[\right)\right)$, so the sets

$\left(f + g\right)^{-1}\left(]\alpha, \infty[\right)$ and $\left(f + g\right)^{-1}\left(]-\infty, \alpha[\right)$ are open.∎

Definition 2.3.7. An homeomorphism from the space (X,τ) onto the space (Y,σ) is a bijection $f : X \longrightarrow Y$ such that f and its inverse f^{-1} are both continuous. For example the function $f : \mathbb{R} \longrightarrow]-1,1[$ given by $f(x) = \dfrac{x}{1+|x|}$ is an homeomorphism with $f^{-1} :]-1,1[\longrightarrow \mathbb{R}$ given by $f^{-1}(x) = \dfrac{x}{1-|x|}$.

Definition 2.3.8. A function $f : X \longrightarrow Y$ is said to be open if the image $f(G)$ of any open set G of X is open in Y. The function is said to be closed if the image $f(F)$ of any closed set F of X is closed in Y.

Proposition 2.3.9. Let $f : X \longrightarrow Y$ be a bijection. The following properties are equivalent:

    (a) f is an homeomorphism from the space (X,τ) onto the space (Y,σ).

    (b) The function f is continuous and open.

    (c) The function f is continuous and closed.

If τ,ω are two topologies on X then $\tau = \omega$ iff the identity mapping $I_X : X \longrightarrow X$ is an homeomorphism.

Proof: Left as an exercise.∎

Remark 2.3.10. (a) A property of a space Y which is inherited from a space X by homeomorphism is called a topological invariant. As an example, consider the following property of X :

(S) $\forall x, y \in X$, if $x \neq y$ there is an open neighbohood U of x and an

      open neighbohood V of y such that $U \cap V = \emptyset$

Now let $f : X \longrightarrow Y$ be an homeomorphism. We prove that the space Y has property (S). Let $u, v \in Y$ with $u \neq v$, there is $x, y \in X$, with $x \neq y$ and $u = f(x)$, $v = f(y)$; let U, V be open sets in X such that $x \in U, y \in V$ and $U \cap V = \emptyset$. Then $f(U), f(V)$ are open in Y and satisfy $u \in f(U)$, $v \in f(V)$, $f(U) \cap f(V) = \emptyset$, so property (S) is a topological invariant.

The property (S) will be studied in details later.

(b) We have to emphasize that not every property of a space can be tranfered to another space by homeomorphism, as is shown by the following example: let $f : [0,1[\longrightarrow \mathbb{R}$, be given by $f(x) = \dfrac{x}{1-x}$, where $[0,1[$ is endowed with the trace topology of $\mathbb{R}$, then it is clear that f is an homeomorphism with inverse $f^{-1}(x) = \dfrac{x}{1+x}$. Now consider the property of Cauchy for sequences of numbers. This property is not preserved by the homeomorphism f since $x_n = 1 - \dfrac{1}{n}$ is Cauchy in $[0,1[$, but $f(x_n) = n - 1$ is not Cauchy in $\mathbb{R}$.

Theorem 2.3.11. (Topology generated by a family of functions)

    Let $\{X_\alpha, \tau_\alpha, \alpha \in A\}$ be a family of topological spaces and let X be a set. For each $\alpha \in A$, let $f_\alpha : X \longrightarrow X_\alpha$ be a function. Then there exists on X a topology denoted by $\tau(f_\alpha, \alpha \in A)$, such that each function f_α is continuous from the space $X, \tau(f_\alpha, \alpha \in A)$ into the space X_α, τ_α. Moreover $\tau(f_\alpha, \alpha \in A)$ is the smallest topology on X with this property. It is called the topology generated by the family of functions $\{f_\alpha, \alpha \in A\}$.

Proof: Put $\mathcal{F} = \{f_\alpha^{-1}(B_\alpha), B_\alpha \in \tau_\alpha, \alpha \in A\}$. The family $\mathcal{F}$ generates the topology $\tau(\mathcal{F})$ for which it is a subbase (Theorem **2.1.12**). We prove that the topology $\tau(\mathcal{F})$ is the needed topology. First each function f_α is continuous from $(X, \tau(\mathcal{F}))$

into (X_α, τ_α) since $C \subset \mathcal{F} \subset \tau(\mathcal{F})$. On the other hand if σ is a topology on X making all the functions f_α continuous, we have
$f_\alpha^{-1}(\tau_\alpha) \subset \sigma, \forall \alpha \in A$, so we deduce $\tau(\mathcal{F}) \subset \sigma$.$\blacksquare$

Proposition 2.3.12. Let g be a function from a space (Y, σ) into the space $(X, \tau(f_\alpha, \alpha \in A))$, then g is continuous if and only if $f_\alpha \circ g : Y \longrightarrow X_\alpha$ is continuous for every $\alpha \in A$.

Proof: If g is continuous, then $f_\alpha \circ g$, as a composition of continuous functions, is continuous for every $\alpha \in A$. Conversely suppose that $f_\alpha \circ g$ is continuous for every α, this implies $g^{-1}\left(f_\alpha^{-1}(\tau_\alpha)\right) \subset \sigma$ for each α and then $g^{-1}(\mathcal{F}) \subset \sigma$, where $\mathcal{F}$ is the generating family of $\tau(f_\alpha, \alpha \in A)$. Consequently if $B \in \tau(f_\alpha, \alpha \in A)$, we can write $B = \bigcup_{\lambda \in \Lambda} I_\lambda$, where each I_λ is a finite intersection of elements in $\mathcal{F}$ (see Theorem **2.1.12**). From the fact that σ is a topology, we deduce easily that $g^{-1}(\tau(f_\alpha, \alpha \in A)) \subset \sigma$ and g is continuous.$\blacksquare$

Proposition 2.3.13. Let $\{X_\alpha, \tau_\alpha, \alpha \in A\}$ be a family of topological spaces and form the cartesian product $X = \Pi_\alpha X_\alpha$ of the sets X_α (see Definition **1.2.7**). Let $\pi_\beta : X \longrightarrow X_\beta$ be the β-coordinate projection given by $\pi_\beta((x_\alpha)) = x_\beta$. Then there is a smallest topology on X making all the projections π_β continuous. Such topology, denoted by $\underset{\alpha}{\otimes}\tau_\alpha$, is called the direct product of the topologies $\tau_\alpha, \alpha \in A$.

Moreover, a function g from a space (Y, σ) into the space $\left(X, \underset{\alpha}{\otimes}\tau_\alpha\right)$ is continuous iff the composition $\pi_\alpha \circ g : Y \longrightarrow X_\alpha$ is continuous for every $\alpha \in A$.

Proof: It is a straightforward application of Theorem **2.3.11** and Proposition **2.3.12** with $X = \Pi_\alpha X_\alpha$ and $f_\alpha = \pi_\alpha, \forall \alpha$.$\blacksquare$

Remark 2.3.14. (a) For $B_\alpha \in \tau_\alpha$ with $\alpha \in \{\alpha_1, \alpha_2, ..., \alpha_n\}$, the set
$\pi_{\alpha_1}^{-1}(B_{\alpha_1}) \cap \pi_{\alpha_2}^{-1}(B_{\alpha_2}) \cap ... \cap \pi_{\alpha_n}^{-1}(B_{\alpha_n})$ is called elementary open set of $\left(\Pi_\alpha X_\alpha, \underset{\alpha}{\otimes}\tau_\alpha\right)$.
One can check that the elementary sets form a base for $\underset{\alpha}{\otimes}\tau_\alpha$, and so every open set in $\underset{\alpha}{\otimes}\tau_\alpha$ contains an elementary open set. Let us observe that
$$\pi_{\alpha_1}^{-1}(B_{\alpha_1}) \cap \pi_{\alpha_2}^{-1}(B_{\alpha_2}) \cap ... \cap \pi_{\alpha_n}^{-1}(B_{\alpha_n}) = B_{\alpha_1} \times B_{\alpha_2} \times ... \times B_{\alpha_n} \times \underset{\alpha \notin \{\alpha_1, \alpha_2, ..., \alpha_n\}}{\Pi} X_\alpha.$$

(b) If $A = \{1, 2, ..., n\}$, all the elementary open sets of $\left(\overset{n}{\underset{\alpha=1}{\Pi}} X_\alpha, \overset{n}{\underset{\alpha=1}{\otimes}} \tau_\alpha\right)$ are of the form $B_1 \times B_2 \times ... \times B_n$, $B_i \in \tau_i$.
As an example it is easy to see that the euclidean topology of $\mathbb{R}^n$ is $\tau \otimes \tau \otimes ... \otimes \tau$, where τ is the euclidean topology of $\mathbb{R}$.

4. Exercises

21. Let (X, τ) be the topological space of Exercise **15**. Prove that a function $f : X \longrightarrow X$ is continuous if and only if f preserves the order in X, that is:
$x \prec y \implies f(x) \prec f(y)$.

22. Let X, τ be a topological space. If R is an equivalence relation on the set X, with quotient X/R, and if $p : X \longrightarrow X/R$ is the canonical mapping (see Definition **1.4.2**), the family $\sigma_R = \left\{B \subset X/R : p^{-1}(B) \in \tau\right\}$ is a topology on X/R (Proposition **2.3.1.**(b)), and p is continuous from X, τ into $X/R, \sigma_R$.

(a) Prove that σ_R is the smallest topology on X/R making p continuous from X, τ into $X/R, \sigma_R$.

(b) Prove that p is an open function iff $p^{-1}(p(U))$ is open in X for every open set $U \subset X$.

23. Let $\{X_\alpha, \tau_\alpha, \alpha \in A\}$ be a family of topological spaces and for each α let $f_\alpha : X \longrightarrow X_\alpha$ be a function from a space (X, τ) into (X_α, τ_α). Define the function $f : X \longrightarrow \underset{\alpha}{\Pi} X_\alpha$ by $f(x) = (f_\alpha(x))$. Prove that f is continuous from (X, τ) into $\left(\underset{\alpha}{\Pi} X_\alpha, \underset{\alpha}{\otimes} \tau_\alpha \right)$ if and only if each function f_α is continuous from (X, τ) into (X_α, τ_α).

5. Separation Axioms

This section deals mainly with two important classes of topological spaces which are the Hausdorff spaces and the normal spaces. These spaces are frequently used in many applications.

Definition 2.5.1. (Hausdorff space)
A topological space (X, τ) is a Hausdorff space if it satisfies the following separation axiom:
For all $x, y \in X$ with $x \neq y$, there is an open neighborhood $V(x)$ of x and an open neighborhood $V(y)$ of y such that $V(x) \cap V(y) = \emptyset$.

Examples 2.5.2. (a) The space $\mathbb{R}$, with the euclidean topology is a Hausdorff space. Indeed if $x \neq y$, put $d = \frac{1}{2}|x - y|$ and take $0 < h < d$, then the open intervals $V(x) =]x - h, x + h[$ and $V(y) =]y - h, y + h[$ satisfy $V(x) \cap V(y) = \emptyset$. likewise we can prove that the space $\mathbb{R}^n$ is a Hausdorff space.

(b) Every discrete topological space (X, τ_1) is a Hausdorff space.

(c) Let X be an infinite set endowed with the topology $\tau = \{\emptyset, A \subset X : A^c \text{ finite}\}$, then (X, τ) is not a Hausdorff space.

Proposition 2.5.3. The following properties are equivalent:
(a) (X, τ) is a Hausdorff space.
(b) For every $x \in X$, if $y \neq x$ there is an an open neighborhood $V(x)$ of x such that $y \notin \overline{V(x)}$.
(c) For each $x \in X$, we have $x = \underset{V \in \mathcal{V}(x)}{\cap} \overline{V}$, where $\mathcal{V}(x)$ is the family of open neighborhoods of x.
(d) The diagonal set $\Delta = \{(x, x) : x \in X\}$ is closed in the product space $X \times X, \tau \otimes \tau$.

Proof: (a) $\Longrightarrow$ (b) if $y \neq x$ there is an open neighborhood $V(x)$ of x and an open neighborhood $V(y)$ of y such that $V(x) \cap V(y) = \emptyset$. So $V(x) \subset X \backslash V(y)$ and then $\overline{V(x)} \subset X \backslash V(y)$ since $X \backslash V(y)$ is closed; this yields $y \notin \overline{V(x)}$.

(b) $\Longrightarrow$ (c) let $x \in X$, by (b) for each $y \neq x$ there $V(x) \in \mathcal{V}(x)$ such that $y \notin \overline{V(x)}$, but then $y \notin \underset{V \in \mathcal{V}(x)}{\cap} \overline{V}$ for all $y \neq x$, this proves that $\underset{V \in \mathcal{V}(x)}{\cap} \overline{V}$ reduces to the point x only.

(c) $\Longrightarrow$ (d) we prove that Δ^c is open. Let $(x, y) \in \Delta^c$, so $y \neq x$; as $x = \underset{V \in \mathcal{V}(x)}{\cap} \overline{V}$, there is an open neighborhood $V(x)$ of x such that $y \in X \backslash \overline{V(x)}$, then put $V(y) =$

$X \backslash \overline{V(x)}$. We get an open neighborhood $V(y)$ of y such that $V(x) \cap V(y) = \emptyset$. Consequently we have $(x, y) \in V(x) \times V(y) \subset \Delta^c$, this shows that Δ^c is open.

$(d) \implies (a)$ if $x \neq y$ then $(x, y) \in \Delta^c$ and there are open sets U, V in X such that: $U \times V \in \tau \otimes \tau, U \times V \subset \Delta^c$ and $(x, y) \in U \times V$; but then $x \in U, y \in V$ and $U \cap V = \emptyset$, whence (a).∎

Proposition 2.5.4. Let $f : X \longrightarrow Y$ be an homeomorphism between (X, τ) and (Y, σ). Then (Y, σ) is Hausdorff if and only if (X, τ) is Hausdorff.

Proof: Left to the reader.∎

Theorem 2.5.5. (a) Every subspace of a Hausdorff space is Hausdorff.

(b) The product space $\left(\Pi_\alpha X_\alpha, \otimes_\alpha \tau_\alpha \right)$ is Hausdorff if and only if each factor (X_α, τ_α) is Hausdorff.

Proof: (a) Let (A, τ_A) be a subspace of a Hausdorff space (X, τ) and let x, y be in A with $x \neq y$. Since (X, τ) is Hausdorff, there is $V(x)$ and $V(y)$ such that $V(x) \cap V(y) = \emptyset$. But $V(x) \cap A$ and $V(y) \cap A$ are in τ_A and disjoint, so (A, τ_A) is Hausdorff.

(b) Suppose (X_α, τ_α) Hausdorff for each α and let $x = (x_\alpha), y = (y_\alpha)$ be in $\Pi_\alpha X_\alpha$ with $x \neq y$, so $x_\alpha \neq y_\alpha$ for some α. Since (X_α, τ_α) is Hausdorff, there are open sets $V(x_\alpha), V(y_\alpha)$ such that $V(x_\alpha) \cap V(y_\alpha) = \emptyset$. By the continuity of π_α, this implies that $\pi_\alpha^{-1}(V(x_\alpha))$ and $\pi_\alpha^{-1}(V(y_\alpha))$ are disjoint open neighborhoods of x and y respectively. This shows that $\left(\Pi_\alpha X_\alpha, \otimes_\alpha \tau_\alpha \right)$ is Hausdorff.

Conversely, let $z = (z_\alpha)$ be fixed in $\Pi_\alpha X_\alpha$ and consider the set

$$Z(\alpha) = \left\{ x = (x_\beta) \in \Pi_\alpha X_\alpha : x_\beta \in X_\beta, x_\beta = z_\beta \ \forall \beta \neq \alpha \right\},$$ endowed with the trace topology of $\otimes_\alpha \tau_\alpha$. Then it easy to check that $Z(\alpha)$ is homeomorphic to the space (X_α, τ_α); since $Z(\alpha)$ is Hausdorff by part (a) of the theorem, we deduce from Proposition **2.5.4** that (X_α, τ_α) is Hausdorff.∎

Theorem 2.5.6. In a Hausdorff space (X, τ) we have:
(a) Every finite set is closed.
(b) Let x be a limit point of a subset $A \subset X$, then for every open neighborhood V of x the set $V \cap A$ is infinite.

Proof: (a) From Proposition **2.5.3** (c) each point of X is closed.
(b) Suppose there is an open neighborhood V of x such that $V \cap (A \backslash x) = \{x_1, x_2, ..., x_n\}$, finite. By (a) the set $\{x_1, x_2, ..., x_n\}^c$ is open and contains x; this implies that $W = V \cap \{x_1, x_2, ..., x_n\}^c$ is an open neighborhood of x such that $W \cap (A \backslash x) = \emptyset$, this is a contradiction with the fact that x is a limit point of A (see Definition **2.1.23**).∎

Now we consider a separation property more severe than the Hausdorff one:
Definition 2.5.7. (Normal space)
A topological space (X, τ) is said to be normal if it is a Hausdorff space satisfying the following property:
For any closed sets F, G such that $F \cap G = \emptyset$, there are open sets U, V with
$\quad F \subset U, G \subset V$ and $U \cap V = \emptyset$.

Examples 2.5.8. (a) Every discrete space (X, τ_1) is normal.

(b) The spaces $\mathbb{R}$ and $\mathbb{R}^n$ endowed with the euclidean topology are normal and more generally we will see in chapter 3 that every metric space is normal.

Theorem 2.5.9. In a topological space (X, τ)
the following properties are equivalent:
(a) The space X is normal
(b) Let F be closed and U open such that $F \subset U$, then there is an open set V
   such that $F \subset V \subset \overline{V} \subset U$.

Proof: (a) $\Longrightarrow$ (b) If F is closed and U open such that $F \subset U$, then U^c is closed and $F \cap U^c = \emptyset$. By (a) there are open sets V, W such that $F \subset V$, $U^c \subset W$, and $V \cap W = \emptyset$. Then we have $V \subset W^c$ and, as W^c is closed, also $\overline{V} \subset W^c$, so we deduce that $F \subset V \subset \overline{V} \subset W^c \subset U$, this proves ($b$).
(b) $\Longrightarrow$ (a) Let F, G be closed such that $F \cap G = \emptyset$, so $F \subset G^c$ and G^c is open. By (b) there is an open set U such that $F \subset U \subset \overline{U} \subset G^c$. But the complement $V = X \backslash \overline{U}$ of $\overline{U}$ is open, contains G and satisfies $U \cap V = \emptyset$, so X is normal.∎

Theorem 2.5.11. (a) Let $f : X \longrightarrow Y$ be an homeomorphism between (X, τ) and (Y, σ). Then (Y, σ) is normal if and only if (X, τ) is normal.
(b) Every closed subspace (A, τ_A) of a normal space (X, τ) is normal.

Proof: (a) straightforward.
(b) Let (X, τ) be normal and let (A, τ_A) be a closed subspace. Let F, G, be closed sets in A with $F \cap G = \emptyset$. Then since A is closed, F, G, are closed in X, so by normality of X there are open sets U, V in X such that $F \subset U, G \subset V$ and $U \cap V = \emptyset$. Now put $U_1 = U \cap A, V_1 = V \cap A$; then U_1, V_1 are open in A and satisfy $F \subset U_1, G \subset V_1$ and $U_1 \cap V_1 = \emptyset$, this proves that (A, τ_A) is normal.∎

The importance of the normal spaces is undoubtedly due to the following results whose proofs can be found in [2].

Theorem 2.5.12.   (Urysohn Lemma)
Let (X, τ) be normal space, then for every disjoint closed sets F, G, there is a continuous function $f : X \longrightarrow \mathbb{R}$ such that :
   (1) $0 \leq f(x) \leq 1$, for all $x \in X$.
   (2) $f(x) = 0$, for all $x \in F$.
   (3) $f(x) = 1$, for all $x \in G$

Theorem 2.5.13.   (Tietze)
Let (X, τ) be normal space and let (A, τ_A) be a closed subspace. Then any continuous function $f : A \longrightarrow \mathbb{R}$ can be extended to a continuous function on all over X. In other words there is a continuous function $g : X \longrightarrow \mathbb{R}$ such that $g(x) = f(x), \forall x \in A$.

6. Exercises

24. Consider the following separation axioms on a space X:
   T_1 : If $x \neq y$, there is an open set containing y but not x.
   T_2 : Hausdorff axiom (Definition **2.5.1**).
   T_3 : T_1 and the following:
   If F is closed and $x \notin F$, there are disjoint open sets U, V
   with $x \in U$ and $F \subset V$.
   T_4 : T_1 and the following:

For any disjoint closed sets F, G, there are disjoint open sets U, V with $F \subset U, G \subset V$.

(a) Prove that X satisties T_1 iff each x is closed.

(b) Prove that $T_4 \implies T_3 \implies T_2 \implies T_1$.

(c) Let X be an infinite set with the topology $\tau = \{\emptyset; A \subset X : A^c \text{ finite}\}$. Prove that (X, τ) is not Hausdorff and it satisfies the axiom T_1.

25. Let (X, τ) and (Y, σ) be topological spaces with (Y, σ) Hausdorff and $f, g : X \longrightarrow Y$ be continuous functions.

(a) Put on $Y \times Y$ the product topology $\sigma \otimes \sigma$ and prove that the function $\varphi : X \longrightarrow Y \times Y$ given by $\varphi(x) = (f(x), g(x))$ is continuous and deduce that the set $\{x \in X : f(x) = g(x)\}$ is closed in X.

(b) Let D be a set dense in X. Prove : $f = g$ on $D \implies f = g$ on X.

(c) Consider the function $h : X \times Y \longrightarrow Y \times Y$, given by $h(x) = (f(x), y)$ and prove that the graph $\Gamma = \{(x, y) \in X \times Y : f(x) = y\}$ of the function f is closed in $X \times Y, \tau \otimes \sigma$.

26. Prove the equivalence of the following properties:

(1) (X, τ) is a normal space

(2) For any disjoint closed sets F, G, there are open sets U, V with $F \subset U, G \subset V$ and $\overline{U} \cap \overline{V} = \emptyset$.

(3) For any disjoint closed sets F, G, there is an open set U with $F \subset U$ and $\overline{U} \cap G = \emptyset$.

27. Let (X, τ) be a normal space, and let $F_1, F_2, ..., F_n$ be closed sets in X with $\overset{n}{\underset{1}{\cap}} F_i = \emptyset$. Prove that there are open sets $V_1, V_2, ..., V_n$ such that $F_i \subset V_i$ and $\overset{n}{\underset{1}{\cap}} \overline{V_i} = \emptyset$. [use Theorem **2.5.9**$(b)$ and induction]

28. Let (X, τ) be a normal space, and let $U_1, U_2, ..., U_n$ be open sets in X with $X = \overset{n}{\underset{1}{\cup}} U_i$. Prove that there are n continuous functions $f_1, f_2, ..., f_n : X \longrightarrow [0, 1]$ such that $\sum_{1}^{n} f_i(x) = 1, \forall x \in X$ and $f_i(x) = 0, \forall x \in X \backslash U_i$, $i = 1, 2, ..., n$.

[Hint: First take open sets $V_1, V_2, ..., V_n$ such that $\overline{V_i} \subset U_i$ and $X = \overset{n}{\underset{1}{\cup}} V_i$ (Exercise **27.**). Next use Urysohn Lemma to get a continuous function g_i associated to the disjoint closed sets $\overline{V_i}, X \backslash U_i$. Then we have $\sum_{1}^{n} g_i(x) > 0 \ \forall x \in X$.

Finally take $f_i(x) = \dfrac{g_i(x)}{\sum_{1}^{n} g_i(x)}$].

7. Connected Spaces

Definition 2.7.1. A topological space (X, τ) is said to be connected if it cannot be written as a union of two non empty disjoint open sets.

A subset $A \subset X$ is connected if the subspace (A, τ_A) is connected. By the definition of the topology τ_A, A is connected if A is not contained in the union of two open setswhose intersections with A are disjoint and non-empty.

Examples 2.7.2. (a) Every discrete space (X, τ_1) containing more than one point is not connected.

(b) Take $X = \{0, 1\}$ and $\tau = \{\emptyset, 0, X\}$, then (X, τ) is connected.

(c) We will see later that the set $\mathbb{R}$ endowed with the usual topology is connected.

(d) The set $\mathbb{Q}$ of rational numbers is not connected. Indeed the open sets of $\mathbb{Q}$
$$U = \left\{x : x > \sqrt{2}\right\} \cap \mathbb{Q}, \ V = \left\{x : x < \sqrt{2}\right\} \cap \mathbb{Q}, \text{ are disjoint and } \mathbb{Q} = U \cup V.$$

Proposition 2.7.3. The following properties are equivalent:

(a) The space (X, τ) connected.

(b) X and $\emptyset$ are the only subsets of X which are open and closed.

(c) Every continuous function from X into the discrete space $\{0, 1\}$ is constant.

Proof: $(a) \implies (b)$ If there is $G \subset X$, G open and closed and $G \neq X, \emptyset$, we would have $X = G \cup G^c$, so X would not be connected.

$(b) \implies (c)$ If there is $f : X \longrightarrow \{0, 1\}$ continuous and not constant, then $f^{-1}(0)$ and $f^{-1}(1)$ are open and closed different from $X, \emptyset$, and satisfy $f^{-1}(0) \cap f^{-1}(1) = \emptyset$, $X = f^{-1}(0) \cup f^{-1}(1)$.

$(b) \implies (c)$ If (X, τ) is not connected, there is U, V disjoint and open with $X = U \cup V$; in this case the function $f : X \longrightarrow \{0, 1\}$ given by $f(x) = 1$ for $x \in U$, and $f(x) = 0$ for $x \in V = U^c$ is continuous and not constant.■

Proposition 2.7.4. Let $f : X \longrightarrow Y$ be an homeomorphism between (X, τ) and (Y, σ). Then (Y, σ) is connected if and only if (X, τ) is connected.

Proof: Left to the reader.■

Theorem 2.7.5. Let $f : X \longrightarrow Y$ be a continuous function from (X, τ) into (Y, σ). If X is connected, then $f(X)$ is connected.

Proof: If $f(X)$ is not connected there would exist a function $g : f(X) \longrightarrow \{0, 1\}$ continuous and not constant. In this case the function $g \circ f : X \longrightarrow \{0, 1\}$ is continuous and not constant, so by Proposition **2.7.3**(c), X would not be connected.■

Theorem 2.7.6. In the space $\mathbb{R}$ the only connected subsets with more than one point are the intervals.

Proof: Let $A \subset \mathbb{R}$ be connected. Suppose that A is not an interval, then there is $a \in A$, $b \in A$, and $c \notin A$ such that $a < c < b$. In this case the sets $A \cap \{x : x < c\}$ and $A \cap \{x : x > c\}$ are open in A disjoint and their union is A, this is a contradiction since A is assumed connected.

Conversely let A be an interval of $\mathbb{R}$. Suppose that A is not connected, then $A = U \cup V$, where U, V are non empty disjoint open sets in (A, τ_A). Let $a \in U, b \in V$, we can assume $a < b$ and form the set $\{x \in U : [a, x] \subset U\}$. This set is not empty since $a \in U$ and it is bounded above by b; consequently $\alpha = Sup \{x \in U : [a, x] \subset U\}$ exists and we have $a \leq \alpha \leq b$. Since A is an interval and since a, b are in A we deduce that $\alpha \in A$. On the other hand, for $\epsilon > 0$ the set $]\alpha - \epsilon, \alpha + \epsilon[\cap A$ is an open neighborhood of α in (A, τ_A), and from the definition of a supremum (Proposition **1.6.3**) we get $\{x \in U : [a, x] \subset U\} \cap \,]\alpha - \epsilon, \alpha + \epsilon[\,\cap A \neq \emptyset$. This impllies that $U \cap \,]\alpha - \epsilon, \alpha + \epsilon[\,\cap A \neq \emptyset$, and then $\alpha \in \overline{U_A}$, where $\overline{U_A}$ is the closure of U in (A, τ_A). Since $U = A \backslash V$, U is also closed in A and then $\overline{U_A} = U$, so we deduce that $\alpha \in U$; since U is open in (A, τ_A), there is an open set G in $\mathbb{R}$ such that $U = G \cap A$. Since A is an interval and since G is a union of open intervals, there is $\epsilon > 0$ such that $]\alpha - \epsilon, \alpha + \epsilon[\,\subset U$, but then we would have $[a, \alpha + \epsilon[\,\subset U$ and this leads to a contradiction with the definition of α. So A is connected and achieves the proof.∎

The following corollary is known as the intermediate value theorem:

Corollary: Let $f : X \longrightarrow \mathbb{R}$ be a continuous function. If X is connected, then for any a, b in $f(X)$ and any c in $[a, b]$ there is $x \in X$ such that $f(x) = c$.

Proof: $f(X)$ is connected in $\mathbb{R}$ by Theorem **2.7.5**, so it is an interval by **2.7.6**, and this yields $[a, b] \subset f(X)$ if a, b are in $f(X)$.∎

Proposition 2.7.7. Let $A \subset X$ be connected, then the closure $\overline{A}$ is connected

Proof: We use Proposition **2.7.3**(c). Let $f : \overline{A} \longrightarrow \{0, 1\}$ be continuous, then the restriction f_A of f to A is continuous, so it is constant. We deduce that f is constant on $\overline{A}$ (see Exercise **25**(b)).∎

Proposition 2.7.8. Let (A_α) be a family of connected subsets in the space X, such that $\underset{\alpha}{\cap} A_\alpha \neq \emptyset$, then $\underset{\alpha}{\cup} A_\alpha$ is connected.

Proof: Let $f : \underset{\alpha}{\cup} A_\alpha \longrightarrow \{0, 1\}$ be continuous, then the restriction f_{A_α} of f to A_α is continuous, so it is constant since A_α is connected. If $x_0 \in \underset{\alpha}{\cap} A_\alpha$, we have $f(x) = f(x_0), \forall x \in A_\alpha$, and this true for each α. This implies that f is constant on $\underset{\alpha}{\cup} A_\alpha$, so $\underset{\alpha}{\cup} A_\alpha$ is connected.∎

In what follows we will see that any not connected space can always be written as a union of connected subsets, called connected components according to the definition:

Definition 2.7.9. Let (X, τ) be a topological space and let $x \in X$. The connected component of x is defined as the union of all connected subsets containing x. It will be denoted by $C(x)$. By Proposition **2.7.8** $C(x)$ is connected and it is the largest connected subset containing x.
A topological space is totally disconnected $C(x) = x$ for all $x \in X$.

Examples 2.7.10. (a) If X is connected, we have $C(x) = X$ for all $x \in X$.
(b) If X is a discrete space, X is totally disconnected.
(c) The set $\mathbb{Q}$ of rational numbers is not discrete, but it is totally disconnected. In fact every connected subset C of $\mathbb{Q}$ reduces to a one point; indeed if such C contains two points $x < y$, then for z irrational with $x < z < y$, the non empty disjoint open sets $U = C \cap \{u \in \mathbb{Q} : u > z\}$, $V = C \cap \{u \in \mathbb{Q} : u < z\}$, would satisfy $C = U \cup V$ contradicting the connectedness of C.

Proposition 2.7.11. $C(x)$ is closed in X for every x.

Proof: By Proposition **2.7.7**, the closure $\overline{C(x)}$ is connected. Since $C(x)$ is maximal we have $\overline{C(x)} \subset C(x)$, so $\overline{C(x)} = C(x)$.∎

Theorem 2.7.12. Let (X, τ) be a topological space
(a) Any two connected components are either equal or disjoint
(b) The binary relation on X given by $xRy \iff y \in C(x)$ is an equivalence relation whose classes are the connected components of X
(c) The quotient space $X/R, \sigma_R$ is totally disconnected (see Example **2.3.3** for the construction of $X/R, \sigma_R$).

Proof: See [].

Definition 2.7.13. A topological space (X, τ) is said to be locally connected if its topology τ has a base formed of open connected sets.

Remark 2.7.14. A topological space can be locally connected without being connected, as is shown by a discrete space. On the other hand a connected space need not be locally connected (see Exercise []). However we have:

Proposition 2.7.15. A topological space (X, τ) is locally connected if and only if the connected components of open sets are open.

Proof: Suppose X locally connected and let $\mathcal{B}$ be a base of open connected sets. Let U be an open set and C a connected component of U. If $x \in C$ then $x \in U$, and since $\mathcal{B}$ is a base there is $B \in \mathcal{B}$ such that $x \in B \subset U$; but C is the connected component of x and B is connected, so we have $x \in B \subset C$, this proves that C is open. Conversely if the connected components of open sets are open, the family $\mathcal{B} = \{C(x), x \in X\}$ is obviously a topological base for X.∎

8. Exercises

29. Let X be an infinite set with the topology $\tau = \{\emptyset; A \subset X : A^c \text{ finite}\}$.
    Prove that (X, τ) is connected.

30. Let $A \subset X$ be a connected subset of X. If the subset B satisfies $A \subset B \subset \overline{A}$, prove that B is connected.

31. Let $f : X \longrightarrow Y$ be a continuous function from (X, τ) into (Y, σ). Suppose X connected and Y totally disconnected. Prove that f is constant
    (by Theorem **2.7.5** $f(X)$ is connected.)

31. Let $f : X \longrightarrow Y$ be a continuous function from (X, τ) into (Y, σ). Prove that $f(C(x)) \subset C(f(x)), \forall x \in X$. If f is an homeomorphism, prove that f defines a one-to-one correspondence between the connected components of X and the connected components of Y.

32. A topological space (X, τ) is said path connected or pathwise connected if for every $a, b \in X$ there is a continuous function $f : [0, 1] \longrightarrow X$ such that
$f(0) = a, f(1) = b$ (we say that a, b are connected by a continuous path). Prove that every path connected space is connected [Hint: Fix $a \in X$, then each $x \in X$ connected to a by a continuous path $\gamma_{ax} = f([0, 1])$, which is a connected subset of X by Theorem **2.7.5**. The conclusion comes from the fact that $X = \bigcup_{x \in X} \gamma_{ax}$ and
Proposition **2.7.8**].

33. In the space $\mathbb{R}^2$ endowed with the euclidean topology, let Γ be the set given by
$$\Gamma = \left\{ (x, y) : 0 < x \leq 1, y = \sin \frac{1}{x} \right\}$$

(a) Prove that Γ is connected (consider the continuous function $\varphi : \,]0,1] \longrightarrow \mathbb{R}^2, \varphi(x) = \left(x, \sin\dfrac{1}{x}\right)$).

(b) Prove that $\overline{\Gamma} = \Gamma \cup (\{0\} \times [-1,1])$

(c) Prove that $\overline{\Gamma}$ is connected but not locally connected (If G is the open set of $\overline{\Gamma}$ given by $G = \overline{\Gamma} \cap \left(\mathbb{R} \times \left]-\dfrac{1}{2}, \dfrac{1}{2}\right[\right)$, then the connected component of the point $x = \left(0, \dfrac{1}{4}\right)$ of G is $\{0\} \times \left]-\dfrac{1}{2}, \dfrac{1}{2}\right[$, which is not open in $\overline{\Gamma}$).

34. A topological space may be connected but not pathwise connected

(In the set $\overline{\Gamma}$ of Exercise **33**, the points $(0,0)$ and $\left(0, \dfrac{1}{\pi}\right)$ are not connected by a continuous path).

CHAPTER 3

Metric Spaces

1. Metrics

This chapter may be considered as a motivation to the study of the topological structures presented in chapter 2. Metric spaces contain convenient concepts which are suitable for the treatment of convergent sequences and continuous functions.

Definition 3.1.1. Let X be a non empty set; a metric (or distance) on X is a function $d : X \times X \longrightarrow \mathbb{R}$ which satisfies the following conditions:$\forall x, y, z$ in X

    (a) $d(x, y) \geq 0$, $d(x, y) = 0 \iff x = y$
    (b) $d(x, y) = d(y, x)$, (symmetry)
    (c) $d(x, y) \leq d(x, z) + d(z, y)$, (triangle inequality)

The set X endowed with a metric d is called a metric space and will be denoted by (X, d).

Examples 3.1.2. (a) Let X be a non empty set, define d by $d(x, x) = 0$ and $d(x, y) = 1$ if $x \neq y$, d is the metric called the discrete metric.
(b) If $X = \mathbb{R}^n$, the euclidean metric is defined by:

$$d(x, y) = \sqrt{\sum_{1}^{n}(x_i - y_i)^2}, \ x, y \in \mathbb{R}^n, \ x = (x_1, ..., x_n), (y_1, ..., y_n)$$

If $n = 1$ we get $d(x, y) = |x - y|$, which is the usual metric of $\mathbb{R}$.
To see triangle inequality, let $x = (x_1, ..., x_n)$, $y = (y_1, ..., y_n)$, $z = (z_1, ..., z_n)$ be in $\mathbb{R}^n$, write: $(x_i - y_i) = (x_i - z_i) + (z_i - y_i)$ then use the inequality

$$\sqrt{\sum_{1}^{n}(\alpha_i + \beta_i)^2} \leq \sqrt{\sum_{1}^{n}\alpha_i^2} + \sqrt{\sum_{1}^{n}\beta_i^2}$$

valid for every finite sequences of real numbers $\alpha_1, ..., \alpha_n$ and $\beta_1, ..., \beta_n$.
(c) Let E be a non empty set and consider the set $B(E)$ of all continuous functions $f : E \longrightarrow \mathbb{R}$, such that $\sup_{x \in X}|f(x)| < \infty$. It is easy to check that $B(E)$ is a vector space on $\mathbb{R}$. If $f, g \in B(E)$, put $d(f, g) = \sup_{x \in X}|f(x) - g(x)|$, then it is straightforward that d defines a metric on $B(E)$.

(d) Let $C[0, 1]$ be the vector space of all continuous functions $f : [0, 1] \longrightarrow \mathbb{R}$. If $f, g \in C[0, 1]$, put $d(f, g) = \int_0^1 |f - g|$ (Riemann integral), then using the properties of Riemann integral it is easy to check that $d(f, g)$ is a metric on $C[0, 1]$.

Definition 3.1.3. Let (X, d) be a metric space. If $x \in X$ and $\alpha > 0$, we define the open ball centered at x and with radius α by the set:

$$B(x, \alpha) = \{y \in X : d(x, y) < \alpha\}$$

Likewise we define the closed ball centered at x with radius α by the set:
$$C\left(x,\alpha\right)=\{y\in X:\ d\left(x,y\right)\leq\alpha\}$$
The sphere centered at x with radius α is the set:
$$S\left(x,\alpha\right)=\{y\in X:\ d\left(x,y\right)=\alpha\}$$

The family of all open balls in a metric space (X,d) constitutes a base of a topology on X according to:

Theorem 3.1.4. Let (X,d) be a metric space.
Let $\mathcal{B}=\{B\left(x,\alpha\right),x\in X,\alpha>0\}$ be the family of all open balls of X.Then $\mathcal{B}$ is a base for the topology $\tau\left(\mathcal{B}\right)$ generated by $\mathcal{B}$.

Proof: We already know that $\mathcal{B}$ is a subbase for the topology $\tau\left(\mathcal{B}\right)$ (see Chap. 2, Definition **2.1.13**). To prove that it is indeed a base, we prove that it satisfies the condition of Theorem **2.1.14**, Chap. 2. To do this let $B\left(x,\alpha\right),B\left(x,\beta\right)$ be open balls with non empty intersection and let $z\in B\left(x,\alpha\right)\cap B\left(x,\beta\right)$; put $\rho=Min\left(\alpha-d\left(x,z\right),\beta-d\left(y,z\right)\right)$, then we have $B\left(z,\rho\right)\subset B\left(x,\alpha\right)\cap B\left(x,\beta\right)$. Indeed if $u\in B\left(z,\rho\right)$, $d\left(z,u\right)<\rho$ and $d\left(x,u\right)\leq d\left(x,z\right)+d\left(z,u\right)<d\left(x,z\right)+\alpha-d\left(x,z\right)$, so $d\left(x,u\right)<\alpha$ and $u\in B\left(x,\alpha\right)$.
Likewise $d\left(y,u\right)\leq d\left(y,z\right)+d\left(z,u\right)<d\left(y,z\right)+\beta-d\left(y,z\right)<\beta$ and then $u\in B\left(y,\beta\right)$; this proves that $B\left(z,\rho\right)\subset B\left(x,\alpha\right)\cap B\left(x,\beta\right)$. So $\mathcal{B}$ is a base for the topology $\tau\left(\mathcal{B}\right)$.$\blacksquare$
In the sequel we denote the topology $\tau\left(\mathcal{B}\right)$ by $\tau\left(d\right)$.

Corollary: $G\in\tau\left(d\right)\iff\forall x\in G\ \exists\alpha>0:B\left(x,\alpha\right)\subset G.$

Definition 3.1.5. A topological space (X,τ) is metrizable if there is a metric d on X such that $\tau=\tau\left(d\right).$

Definition 3.1.6. The metrics d and ρ on X are equivalent if $\tau\left(d\right)=\tau\left(\rho\right).$
Let us point out that the definition does not signify d and ρ determine the same open balls. However we have:

Proposition 3.1.7. The metrics d and ρ on X are equivalent if and only if for each $x\in X$ and each $\epsilon>0$:
$\quad$(1) $\exists\alpha_1=\alpha_1\left(x,\epsilon\right)>0:\rho\left(x,y\right)<\alpha_1\implies d\left(x,y\right)<\epsilon$
$\quad$(2) $\exists\alpha_2=\alpha_2\left(x,\epsilon\right)>0:d\left(x,y\right)<\alpha_2\implies\rho\left(x,y\right)<\epsilon$

Proof: We know from Proposition **2.3.9** of Chap. 2 that $\tau\left(d\right)=\tau\left(\rho\right)$ iff the identity mapping $I_X:X\longrightarrow X$ is an homeomorphism. Properties $\left(1\right),\left(2\right)$ are the expression of this fact (see sction 3 below about continuity in metric spaces).$\blacksquare$

Proposition 3.1.8. In $\mathbb{R}^n$, the following formulas define two metrics equivalent to the euclidean metric:

$$\delta\left(x,y\right)=\underset{1\leq i\leq n}{Max}\ |x_i-y_i|$$

$$\gamma\left(x,y\right)=\sum_1^n|x_i-y_i|$$

Proof: First it is clear that δ,γ are metrics on $\mathbb{R}^n$. The equivalence property results easily from the following inequalities:

$$\frac{1}{\sqrt{n}} \underset{1 \le i \le n}{Max}\, a_i \le \frac{1}{\sqrt{n}} \sqrt{\sum_{1}^{n} a_i^2} \le \underset{1 \le i \le n}{Max}\, a_i$$

$$\underset{1 \le i \le n}{Max}\, a_i \le \sum_{1}^{n} a_i \le n \underset{1 \le i \le n}{Max}\, a_i$$

which are valid for all positive numbers $a_1, ..., a_n$. $\blacksquare$

When a metric space (X, d) is equipped with the topology $\tau(d)$, all the concepts introduced in chapter 2, like closure, interior, limit point..., are well defined in $(X, \tau(d))$. As an example, let us prove that the closed ball $C = C(x, \alpha)$ is a closed set in $(X, \tau(d))$. Let $y \in X \backslash C$ then $d(x, y) > \alpha$ and if $0 < \beta < d(x, y) - \alpha$, we get by the triangle inequality: $d(y, z) < \beta \implies d(x, y) > \alpha$, which proves that $B(y, \beta) \subset X \backslash C$, so $X \backslash C$, is open by the corollary of Theorem **3.1.4.**

Proposition 3.1.9. The topological space $(X, \tau(d))$, is Hausdorff.

Proof: Let $x, y \in X, x \neq y$; take $0 < \alpha < \frac{1}{2} d(x, y)$, then the open balls $B(x, \alpha)$ and $B(y, \alpha)$ are two disjoint open neighborhoods of x and y respectively. $\blacksquare$

Proposition 3.1.10. Let $A \subset X$ be a subset of a metric space (X, d) and let x be a limit point of A. Then for every $\alpha > 0$ the intersection $B(x, \alpha) \cap A$ is infinite.

Proof: Apply Definition **2.1.23** Chap.2. $\blacksquare$

Definition 3.1.11. (Metric subspace)
Let $A \subset X$ be a subset of a metric space (X, d). The restriction $d_A : A \times A \longrightarrow \mathbb{R}$ of the metric d to $A \times A$ is obviously a metric on A and the metric space (A, d_A) so defined is a metric subspace of the metric space (X, d).

Proposition 3.1.12. Let (A, d_A) be a metric subspace of the metric space (X, d). If $x \in A$ and $\alpha > 0$, consider the open ball $B_A(x, \alpha)$ in (A, d_A), centred at x with radius α, given by $B_A(x, \alpha) = \{y \in A : d_A(x, y) < \alpha\}$. Then we have $B_A(x, \alpha) = B(x, \alpha) \cap A$, for every $x \in A$ and $\alpha > 0$, where $B(x, \alpha)$ is the open ball in (X, d). Conversely for any open ball $B(x, \alpha)$ in (X, d) the set $B(x, \alpha) \cap A$, if non-empty, is the union of open balls in (A, d_A).

Proof: Left to the reader. $\blacksquare$

Theorem 3.1.13. Let (A, d_A) be a metric subspace of the metric space (X, d). Then the topology $\tau(d_A)$ defined on A by the metric d_A is identical to the trace topology induced by $\tau(d)$ on A.

Proof: Let U_A be an open set in $\tau(d_A)$, then (Theorem **3.1.4.**), $U_A = \underset{x \in U_A}{\cup} B_A(x, \alpha_x)$. From Proposition **3.1.12**, $B_A(x, \alpha_x) = B(x, \alpha_x) \cap A$, so we get $U_A = \underset{x \in U_A}{\cup} B(x, \alpha_x) \cap A = G \cap A$, where $G = \underset{x \in U_A}{\cup} B(x, \alpha_x)$ is open in $(X, \tau(d))$, therefore $U_A \in \tau(d) \cap A = (\tau(d))_A$. Conversely if U is open in the trace topology of $\tau(d)$ on A, $U = G \cap A$, for some $G \in \tau(d)$. But then $G = \underset{x \in G}{\cup} B(x, \alpha_x)$, where the $B(x, \alpha_x)$ are open balls in $(X, \tau(d))$ and this yields $U = \underset{x \in G}{\cup} (B(x, \alpha_x) \cap A)$. From Proposition **3.1.12**, each $B(x, \alpha_x) \cap A$ is the union of open balls in (A, d_A), so U itself is the union of such balls and this gives $U \in \tau(d_A)$. $\blacksquare$

Definition 3.1.14. (a) Let $A \subset X$ be a subset of a metric space (X, d). The diameter of A is defined by $\delta(A) = \sup\limits_{(x,y)\in A\times A} d(x,y)$. The set A is bounded if $\delta(A) < \infty$ and the metric d is bounded if $\delta(X) < \infty$.
If there is $x_0 \in X$ such that $M = \sup\limits_{y\in A} d(x_0, y) < \infty$, A is bounded and $\delta(A) \le 2M$.
(b) The distance from $x \in X$ to the subset A is defined by $d(x, A) = \inf\limits_{y\in A} d(x, y)$.

Proposition 3.1.15. In a metric space (X, d), the formula $\rho(x, y) = \dfrac{d(x, y)}{1 + d(x, y)}$ defines a bounded metric equivalent to d.

Proof: The function $t \longrightarrow \dfrac{t}{1 + t}$ and its inverse $t \longrightarrow \dfrac{t}{1 - t}$ are increasing on their domains and we have $u = \dfrac{t}{1 + t} \iff t = \dfrac{u}{1 - u}$. So $d(x, y) = \dfrac{\rho(x, y)}{1 - \rho(x, y)}$ and Proposition **3.1.7** gives the equivalence of ρ and d.∎

2. Exercises

For all the usual topologial concepts (interior, closure, boundary) the reader is refered to Chapter 2, sections 1,2.

35. Let $\mathbb{R}^2$ be equipped with the euclidean metric. Determine $\overset{o}{A}, \overline{A}, \partial A$, for each of the following sets:
$$L = (0, 0) \cup \{(x, y) : \alpha < d((0, 0), (x, y)) < \beta\}, \text{ for } 0 < \alpha < \beta$$
$$M = \mathbb{Z}^2$$
$$N = \{(x, y) : y \ge x\}$$

36. Let X be a set endowed with the discrete metric d (Example **3.1.2** (a)).
(a) Prove that $\tau(d)$ is the discrete topology on X.
(b) Determine $B(x, \alpha)$ and $C(x, \alpha)$ for $\alpha = 0, \alpha = 1$
Deduce that $C(x, \alpha)$ is not in general the closure of $B(x, \alpha)$

37. Let $\mathbb{R}^n$ be equipped with the euclidean metric.
(a) Prove that the closure of the open ball $B(x, \alpha)$ is the closed ball $C(x, \alpha)$.
(b) Prove that the boundary of $B(x, \alpha)$ is the sphere $S(x, \alpha)$.

38. Let E be a bounded subset of $\mathbb{R}$ with $x = Sup\, E, y = Inf\, E$.
Prove that $x, y \in \overline{E}$.

39. Let $(X_i, d_i), i = 1, ..., n$, be n metric spaces. On the product $X = X_1 \times ... \times X_n$, we define for $x = (x_1, ..., x_n), y = (y_1, ..., y_n)$:

$$d(x, y) = \sqrt{\sum_1^n d_i^2(x_i, y_i)}$$
$$\rho(x, y) = \underset{1\le i\le n}{Max}\, d_i(x_i, y_i)$$
$$\gamma(x, y) = \sum_1^n d_i(x_i, y_i)$$

(a) Prove that these formulas defines three equivalent metrics on the product X.

(b) Prove that $\tau(d) = \tau(\rho) = \tau(\gamma) = \overset{n}{\underset{i}{\otimes}}\tau(d_i)$, where $\overset{n}{\underset{i}{\otimes}}\tau(d_i)$ is the product topology of the topologies $\tau(d_i)$ on X.

40. A pseudometric on a set X is a function $d_0 : X \times X \longrightarrow \mathbb{R}$ which satisfies the following conditions: $\forall x, y, z$ in X

$\quad$ (a) $d_0(x, y) \geq 0$, $d_0(x, x) = 0$

$\quad$ (b) $d_0(x, y) = d_0(y, x)$, (symmetry)

$\quad$ (c) $d_0(x, y) \leq d_0(x, z) + d_0(z, y)$, (triangle inequality)

(1) Prove that the family $\mathcal{B}_0$ of all open balls $B(x, \alpha) = \{y \in X : d_0(x, y) < \alpha\}$ is a base for the topology $\tau(\mathcal{B}_0)$ generated by $\mathcal{B}_0$ and denoted by $\tau(d_0)$.

(2) Prove that $(X, \tau(d_0))$ is Hausdorff if and only if d_0 is a metric.

(3) Define on X the binary relation R by: $xRy \iff d_0(x, y) = 0$

$\quad$ (a) Prove that R is an equivalence relation on X

$\quad$ (b) Define $D : X/R \times X/R \longrightarrow \mathbb{R}$ by the recipe:

$\quad$ $D(C_x, C_y) = d_0(x, y)$, where C_x, C_y are the equivalence classes of x, y.

$\quad$ Prove that D defines unambiguously a metric on X/R.

$\quad$ (c) If X is endowed with topology $\tau(d_0)$, prove that the canonical map

$\quad$ $p : X \longrightarrow X/R$ is continuous , open, and closed.

41. Let (X, d) be a metric space and fix $c > 0$.

(a) Define ρ_c by $\rho_c(x, y) = Min\{c, d(x, y)\}$, $x, y \in X$

Prove that ρ_c is a bounded metric equivalent to d.

(b) Prove that $d(x, A) = 0 \iff x \in \overline{A}$.

3. Cauchy Sequences-Complete Spaces

Definition 3.2.1. Let (X, d) be a metric space and let (x_n) be a sequence in X. We say that:

$\quad$ (a) (x_n) converges to the limit x if $\underset{n}{\lim} d(x_n, x) = 0$.

$\quad$ (b) (x_n) is a Cauchy sequence if $\underset{n,m}{\lim} d(x_n, x_m) = 0$.

Let us point out that:

$\quad$ (a) means: $\forall \epsilon > 0 \; \exists N \geq 1 : \forall n \geq N, \; d(x_n, x) < \epsilon$

$\quad$ (b) means: $\forall \epsilon > 0 \; \exists N \geq 1 : \forall n, m \geq N, \; d(x_n, x_m) < \epsilon$

Proposition 3.2.2. Let (x_n) be a sequence in (X, d), then:

$\quad$ (a) If x_n converges, the limit x is unique.

$\quad$ (b) If x_n converges, x_n is Cauchy.

Proof: (a) If x_n converges to x and to y then for every n we have

$\quad$ $d(x, y) \leq d(x, x_n) + d(x_n, y) \longrightarrow 0, n \longrightarrow \infty$, so $d(x, y) = 0$ and $x = y$.

(b) Suppose $\underset{n}{\lim} x_n = x$ then for every n, m

$\quad$ $d(x_n, x_m) \leq d(x_n, x) + d(x, x_m) \longrightarrow 0, n, m \longrightarrow \infty.\blacksquare$

The converse of point (b) is not true in general: take $X = \,]0, 1]$ with the usual metric and $x_n = \dfrac{1}{n}$, then x_n is Cauchy but does not converge in $]0, 1]$. This observation leads to the following definition.

Definition 3.2.3. A metric space (X, d) is said to be complete if every Cauchy sequence of X converges to a point of X. A subset $A \subset X$ is complete if the metric subspace (A, d_A) is complete.

Example 3.2.4. (a) $\mathbb{R}$ and $\mathbb{R}^n$ are complete for the euclidean metric.
(b) Let E be a non empty set and consider the set $B(E)$ of all continuous functions
$f : E \longrightarrow \mathbb{R}$, such that $\underset{x \in X}{Sup} |f(x)| < \infty$, equipped with the metric
$d(f,g) = \underset{x \in X}{Sup} |f(x) - g(x)|$, (see Example **3.1.2** (c)). $B(E)$ is complete since the
convergence $d(f_n, f) \longrightarrow 0$ with respect to the metric d is equivalent to the uniform
convergence of f_n to f.

Proposition 3.2.5. Let A be a subset of a metric space (X, d). Then A is closed
if and only if for every convergent sequence x_n of A the limit of x_n is in A.

Proof: Suppose A closed and let x_n be a sequence of A convergent to $x \in X$. If
x is in the open set $X \backslash A$, there is $\epsilon > 0$ such that $B(x, \epsilon) \subset X \backslash A$. This implies
that $B(x, \epsilon)$ does not contain any element x_n of the sequence of A, contradicting
the fact that x is the limit of x_n, so we have $x \in A$.
Conversely suppose that for every convergent sequence x_n of A the limit of x_n is
in A. We prove that A is closed by showing that $\overline{A} = A$. If $z \in \overline{A}$, then for every
n the set $Z_n = B\left(z, \dfrac{1}{n}\right) \cap A$ is non-empty. For each n choose z_n in Z_n, then z_n is
a sequence in A converging to z, so $z \in A$ and $\overline{A} = A.$■

Theorem 3.2.6. Let (X, d) be a complete metric space and $A \subset X$.
Then A is complete if and only if A is closed.

Proof: Suppose A complete. If $x \in \overline{A}$, the argument used in the above proof shows
the existence of a sequence x_n in A converging to x. So x_n is Cauchy in A and
since A is assumed complete, we get $x \in A$, this yields $\overline{A} = A$, and A is closed.
Conversely suppose A closed and let x_n be Cauchy in A; then x_n is Cauchy in X,
so it converges to some x in X, since X is complete. By the above proposition
$x \in A$ since A is closed and this proves that A is complete.■

Definition 3.2.7. Let $f : X \longrightarrow Y$ be a function from (X, d) into (Y, ρ), we say
that f is an isometry if $\rho(f(x), f(y)) = d(x, y), \forall x, y \in X.$

Theorem 3.2.8.   **(Completion of a metric space)**
    Let (X, d) be a metric space. There exists a complete metric space (X^*, d^*)
and an isometry $\varphi : X \longrightarrow X^*$ such that $\varphi(X)$ is dense in X^*. Moreover (X^*, d^*)
is unique in the sense that if (Y^*, ρ^*) is an other complete metric space satisfying
the above property, there is an isometry ψ from (X^*, d^*) onto (Y^*, ρ^*).

Proof: See $[]$.■.

Theorem 3.2.9. Let (A_n) be a sequence of closed subsets in a complete
metric space (X, d) such that $A_1 \supset A_2 \supset ...$, and $\delta(A_n) \longrightarrow 0, n \longrightarrow \infty$, where
$\delta(A_n) =$diameter of A_n. Then $\underset{n}{\cap} A_n = \{x\}$ for some $x \in X$.

Proof: For each n choose $x_n \in A_n$; if $m \geq n$ we have $x_n, x_m \in A_n$ and $d(x_n, x_m) \leq$
$\delta(A_n) \longrightarrow 0, n \longrightarrow \infty$, so the sequence x_n is Cauchy in X. Since X is complete x_n
converges to some $x \in X$. Now for every $m \geq 1$, we have $x_n \in A_m, \forall n \geq m$, but the
set A_m is closed, so we deduce that $x \in A_m$ (Proposition **3.2.5**) and then $x \in \underset{n}{\cap} A_n$.
If $z \in \underset{n}{\cap} A_n$ we would have $d(x, z) \leq \delta(A_n), \forall n$, which implies $d(x, z) = 0$ that is
$z = x$, and this shows that $\underset{n}{\cap} A_n = \{x\} .$■

4. Exercises

42. Let x_n be a sequence in a metric space (X, d). Prove that:
x_n Cauchy $\implies$ every subsequence x_{n_k} of x_n is Cauchy
x_n converges to $x \implies$ every subsequence x_{n_k} of x_n converges to x

43. Prove that if x_n is Cauchy in the metric space (X, d), the set
$E = \{x_1, x_2, ...\}$ is bounded, that is $\delta(E) < \infty$.

44. Let x_n be a Cauchy sequence in a metric space (X, d).
Prove that x_n contains a subsequence x_{n_k} satisfying:
$$d\left(x_{n_{k+1}}, x_{n_k}\right) < \frac{1}{2^k}, \text{ for each } k \geq 1$$

45. Let (X, d) be a metric space. Prove that:
(a) The intersection of every family of complete subspaces of (X, d)
is a complete subspace
(b) The union of every finite family of complete subspaces of (X, d)
is a complete subspace

5. Uniformly continuous functions

For topics on continuous functions in general topological spaces, the reader is refered to section 3 of chapter 2. In this section we deal with limit processes in metric spaces.

Definition 3.5.1. Let $f : X \longrightarrow Y$ be a function from (X, d) into (Y, ρ).
We say that f has the limit $y_0 \in Y$ at $x_0 \in X$ if :
$$\forall \epsilon > 0, \exists \alpha = \alpha(x_0, \epsilon) > 0 : x \in X, d(x, x_0) < \alpha \implies \rho(f(x), y_0) < \epsilon$$
which we denote by $\lim_{x \to x_0} f(x) = y_0$.
We say that f is continuous at x_0 if $\lim_{x \to x_0} f(x) = f(x_0)$.

It is useful that this limit process can be handled with sequenences:

Theorem 3.5.2. The following are equivalent:
(a) $\lim_{x \to x_0} f(x) = y_0$
(b) For every sequence x_n in X converging to x_0, the sequence $y_n = f(x_n)$
in Y converges to y_0.

Proof: Assume (a) satisfied, then
$$\forall \epsilon > 0, \exists \alpha > 0 : x \in X, d(x, x_0) < \alpha \implies \rho(f(x), y_0) < \epsilon.$$
Let x_n be convergent to x_0, so there is $N \geq 1$ such that $n \geq N \implies d(x_n, x_0) < \alpha$
and this implies $\rho(f(x_n), y_0) < \epsilon$, that is $f(x_n)$ converges to y_0. This yields (b).
Suppose that (a) is not satisfied then
$$\exists \epsilon > 0 : \forall \alpha > 0, \exists x \text{ with } d(x, x_0) < \alpha \text{ and } \rho(f(x), y_0) \geq \epsilon.$$
Taking $\alpha = \frac{1}{n}$, we get x_n such that $d(x_n, x_0) < \frac{1}{n}$, and $\rho(f(x_n), y_0) \geq \epsilon, \forall n$. This
gives a sequence x_n converging to x_0, without convergence of $f(x_n)$ to y_0, so (b) is
not satisfied.■

Corollary: The function $f : X \longrightarrow Y$ is continuous at $x_0 \in X$ if and only if for
every sequence x_n in X converging to x_0, the sequence $f(x_n)$ converges to $f(x_0)$.

Definition 3.5.3. The $f : X \longrightarrow Y$ be a function from (X, d) into (Y, ρ) is said to be uniformly continuous on X if:
$$\forall \epsilon > 0, \exists \alpha = \alpha(\epsilon) > 0 : \forall x, x' \in X, \, d(x, x') < \alpha \implies \rho(f(x), f(x')) < \epsilon.$$

Examples 3.5.4. (a) For each $z \in X$, the function $x \longrightarrow d(x, z)$ is uniformly continuous on X. Indeed $\forall x, y \in X$ we have:
$$d(x, z) \leq d(x, y) + d(y, z)$$
$$d(y, z) \leq d(y, x) + d(x, z)$$
This yields $|d(x, z) - d(y, z)| \leq d(x, y)$, and the uniform continuity of the function $x \longrightarrow d(x, z)$ is satisfied with $\alpha(\epsilon) = \epsilon$.

(b) More generally for $A \subset X$, the function $x \longrightarrow d(x, A)$, where $d(x, A)$ is the distance between x and A, is uniformly continuous on X. $(d(x, A) = \inf\limits_{y \in A} d(x, y))$.
To see this, take $x, y \in X$ and $z \in A$, then we have:
$$d(x, A) \leq d(x, z) \leq d(x, y) + d(y, z)$$
$$\implies d(x, A) \leq \inf\limits_{z \in A} (d(x, y) + d(y, z)) = d(x, y) + d(y, A)$$
$$d(y, A) \leq d(y, z) \leq d(y, x) + d(x, z)$$
$$\implies d(y, A) \leq \inf\limits_{z \in A} (d(y, x) + d(x, z)) = d(y, x) + d(x, A)$$

This yields $|d(x, A) - d(y, A)| \leq d(x, y)$, and the uniform continuity of the function $x \longrightarrow d(x, A)$ is satisfied with $\alpha(\epsilon) = \epsilon$.

There is a useful class of uniformly continuous functions $f : X \longrightarrow Y$ given by:

Definition 3.5.5. $f : X \to Y$ is called Lipschitz function if there exists a real constant $K \geq 0$ such that, for all $x, x' \in X, \rho(f(x), f(x')) \leq K.d(x, x')$.
Any such K is refereed to as a Lipschitz constant for the function f. if $0 \leq K < 1$ the function is called a contraction.
Uniform continuity of a Lipschitz function is easy to prove.

Note: In the theory of differential equations, Lipschitz continuity is the central condition of the Picard–Lindelof theorem which guarantees the existence and uniqueness of the solution to an initial value problem. The contraction type is used in the Banach fixed point theorem (see Chapter 5).
Uniform continuity is useful in handling Cauchy sequences, contrary to simple continuity (see Remark **2.3.10** (b) Chapter 2).

Proposition 3.5.6. Let $f : X \longrightarrow Y$ be a function from (X, d) into (Y, ρ).
If f is uniformly continuous, then the image by f of a Cauchy sequence in X is a Cauchy sequence in Y.

Proof: Since f is uniformly continuous, we have:
$$\forall \epsilon > 0, \exists \alpha = \alpha(\epsilon) > 0 : \forall x, x' \in X, \, d(x, x') < \alpha \implies \rho(f(x), f(x')) < \epsilon.$$
Let x_n be a Cauchy sequence in X then $\exists N \geq 1 : \forall n, m \geq N, \, d(x_n, x_m) < \alpha$ and this implies $\rho(f(x_n), f(x_m)) < \epsilon$, so $f(x_n)$ is Cauchy in Y. ∎

Corollary: Let $f : X \longrightarrow Y$ be a bijection such that f and f^{-1} are uniformly continuous. Then x_n is Cauchy in X if and only if $f(x_n)$ is Cauchy in Y.
In particular (Y, ρ) is complete if and only if (X, d) is complete.

6. Exercises

46. Prove that for every subset A of a metric space (X, d), we have $\overline{A} = \{x \in X : d(x, A) = 0\}$.

47. For any closed sets F, G in a metric space (X, d) such that $F \cap G = \emptyset$, define the function $f : X \longrightarrow [0, 1]$ by $f(x) = \dfrac{d(x, F)}{d(x, F) + d(x, G)}$. Prove that the sets $U = \{x : d(x, F) < d(x, G)\}, V = \{x : d(x, G) < d(x, F)\}$ are open with $U \cap V = \emptyset$ and $F \subset U, G \subset V$. Deduce that any metric space is normal.

48. (a) Let F be closed in (X, d), put $U_n = \left\{x : d(x, F) < \dfrac{1}{n}\right\}, n \geq 1$.

     Prove that U_n is open for each n and $F = \underset{n}{\cap}\, U_n$.

(b) Let G be open in (X, d), put $V_n = \left\{x : d(x, G) \geq \dfrac{1}{n}\right\}, n \geq 1$.

     Prove that V_n is closed for each n and $G = \underset{n}{\cup}\, V_n$.

49. Let (X, d) be a connected metric space containing more than one point. Prove that X is infinite uncountable.

[Fix $z \in X$, the function $x \longrightarrow d(x, z)$ is not constant and realize a surjection from X onto an interval of $\mathbb{R}$].

50. Let (X, d) be a metric space and put on the product $X \times X$ the metric $\rho((x, y), (x', y')) = d(x, x') + d(y, y')$. Prove that the function $(x, y) \longrightarrow d(x, y)$ is uniformly continuous on the space $(X \times X, \rho)$.

51. Let $f : X \longrightarrow X$ be a contraction of a metric space (X, d) with constant $0 < K < 1$ (see Definition **3.5.5**). Fix x in X and define the sequence x_n by:

     $x_0 = x$ and $x_n = f(x_{n-1})$, for all $n \geq 1$.

(a) Prove that x_n is Cauchy.

(b) If X is complete prove that x_n to an α satisfying $f(\alpha) = \alpha$.

(c) Prove that α is the unique solution of the equation $f(t) = t, t \in X$.

     So α is independent of the initial point x.

7. Countable Bases-Separable Spaces

Definition 3.7.1. A topological space (X, τ) is separable if there is a countable subset $D \subset X$ dense in X. The space X is second countable if it has a countable base that is if the topology τ has a base formed by a sequence of open sets.

Examples 3.7.2. (a) Every countable discrete space is separable
     and has a countable base.

(b) The space $\mathbb{R}$ with the usual topology is separable since the set $\mathbb{Q}$ of
     rationals is countable and dense.

For metric spaces we have the following fundamental:

Theorem 3.7.3. Let (X, d) be a metric space. The following are equivalent:
     (a) X is separable.
     (b) The topology $\tau(d)$ induced by d on X has a countable base.

Proof: Assume (a) and let $D = \{x_1, x_2, ...\}$ be a dense sequence in X. Then the countable family of open balls given by $\left\{B\left(x_n, \dfrac{1}{m}\right), n, m \geq 1\right\}$ is a base for the topology $\tau(d)$. Indeed it is enough to prove that for each $x \in X$ and each $\alpha > 0$ there is $n, m \geq 1$ such that $x \in B\left(x_n, \dfrac{1}{m}\right) \subset B(x, \alpha)$ (see Theorem **3.1.4**). Let $m \geq 1$ such that $\dfrac{1}{m} < \dfrac{\alpha}{2}$, by the density of D there is $n \geq 1$ such that $x_n \in B\left(x, \dfrac{1}{m}\right)$ which gives $x \in B\left(x_n, \dfrac{1}{m}\right)$. On the otherhand if $y \in B\left(x_n, \dfrac{1}{m}\right)$, we have $d(x_n, y) < \dfrac{1}{m}$ and $d(x, y) \leq d(x, x_n) + d(x_n, y) < \dfrac{1}{m} + \dfrac{1}{m} < \alpha$, so $y \in B(x, \alpha)$, that is $B\left(x_n, \dfrac{1}{m}\right) \subset B(x, \alpha)$.

Assume (b) and let $\mathcal{B} = \{B_1, B_2, ...\}$ be a countable base for $\tau(d)$. In each B_n of $\mathcal{B}$ choose x_n, then it is clear that the sequence $\{x_1, x_2, ...\}$ is dense in X.∎

Proposition 3.7.4. Let X, Y be topological spaces and let $f : X \longrightarrow Y$ be a function. Suppose that f is onto, then we have:
(a) If f is continuous and X separable then Y is separable.
(b) If f is continuous open and X has a countable base then
   Y has a countable base.

Proof: (a) Let $D = \{x_1, x_2, ...\}$ be a dense sequence in X. We have
$Y = f(X) = f(\overline{D}) \subset \overline{f(D)}$, where the inclusion comes from the continuity of f. Since $f(D)$ is countable, Y is separable.
(b) Let $\mathcal{B} = \{B_1, B_2, ...\}$ be a countable base for X. Since f is open, $f(B_n)$ is open in Y for every n. We prove that the family $f(\mathcal{B}) = \{f(B_1), f(B_2), ...\}$ is a base for Y. Let V be open in Y and let $y \in V$, then $f^{-1}(V)$ is open in X by the continuity of f. Let $x \in f^{-1}(V)$ such that $y = f(x)$; since $\mathcal{B}$ is a base, there is B_n such that $x \in B_n \subset f^{-1}(V)$, whence $y = f(x) \in f(B_n) \subset f\left(f^{-1}(V)\right)$. But $f\left(f^{-1}(V)\right) = V$ because f is onto, so $f(\mathcal{B})$ is a countable base for Y.∎

We close this section by the following theorem whose proof can be found in [].
For the construction of the product of topological spaces, see Proposition **2.3.13** in chapter 2.

Theorem 3.7.5. Let $\left(\underset{n}{\Pi} X_n, \underset{n}{\otimes} \tau_n\right)$ be the product of a sequence (X_n, τ_n) of topological spaces. Then $\left(\underset{n}{\Pi} X_n, \underset{n}{\otimes} \tau_n\right)$ is separable (resp. has countable base) if and only if each space (X_n, τ_n) is separable (resp. has countable base).

Corollary: The euclidean spaces $\mathbb{R}^n, n \geq 1$ and the infinite product space $\mathbb{R}^{\mathbb{N}}$ are separable and second countable.

As for infinite product of metric spaces we quote the following ingredients:
Let (X_n, d_n) be a sequence of metric spaces with diameters $\delta_n = \delta(X_n)$. On the product $X = \underset{n}{\Pi} X_n$ define the function $\rho : X \times X \longrightarrow \mathbb{R}$ by $\rho(x, y) = \underset{n}{Sup}\, d_n(x_n, y_n)$, where $x = (x_n), y = (y_n)$. Then we have:
Theorem 3.7.6. (a) Suppose that the sequence δ_n is bounded, then ρ defines a metric on X.
(b) The product topology $\underset{n}{\otimes} \tau(d_n)$ on X coincides with the topology $\tau(\rho)$ induced by ρ if and only if $\underset{n}{Lim}\, \delta_n = 0$.

Theorem 3.7.7. Suppose there is a sequence (ϵ_n) of positive numbers such that $\sum_n \epsilon_n \delta_n < \infty$ and define $\sigma : X \times X \longrightarrow \mathbb{R}$ by $\sigma(x,y) = \sum_n \epsilon_n d_n(x_n, y_n)$. Then σ is a metric on X and we have $\tau(\sigma) = \underset{n}{\otimes}\tau(d_n)$.

Proof: see $[]$.

8. Exercises

52. Let (X,τ) be a second countable topological space and let $\mathcal{F} = \{U_\alpha, \alpha \in A\}$ be a family of open sets such that $X = \underset{\alpha \in A}{\cup} U_\alpha$. Prove that $X = \underset{n}{\cup}U_{\alpha_n}$ for some countable subfamily $\{U_{\alpha_n}, n \geq 1\}$ of $\mathcal{F}$ (**Lindelof property**).
[Let $\mathcal{B} = \{B_1, B_2, ...\}$ be a countable base for X. Since each U_α is a union of elements in $\mathcal{B}$, we get a subfamily $\mathcal{S}$ of $\mathcal{B}$ which covers X. Put $\mathcal{S} = \{B_{n_1}, B_{n_2}, ...\}$ and for each n_j take the U_{α_j} containing B_{n_j}].

53. Let (X,d) be a metric space in which every infinite subset has a limit point. By completing the following steps, prove that X is separable.
(a) Let $\delta > 0$ and fix $x_1 \in X$. Define the sequence x_n as follows:
  $x_2 \in X\backslash B_1$ where $B_1 = B(x_1, \delta)$
  $x_3 \in X\backslash(B_1 \cup B_2)$ where $B_2 = B(x_2, \delta)$ and so on.
Prove that $\forall n \neq m, d(x_n, x_m) \geq \delta$ and such sequence is finite.
(b) Deduce that for each $\delta > 0$ the space X can be covered by a finite set of open balls of radius δ.
(c) If $\delta = \dfrac{1}{n}$, let D_n be the finite set of the centers of open balls that cover X
Prove that $D = \underset{n}{\cup}D_n$ is dense in X.

9. Baire Spaces

Definition 3.9.1. A topological space (X,τ) is a Baire space if the intersection of every countable family of dense open sets in X is a dense subset of X.

Proposition 3.9.2. Let X be a Baire space and let $\{F_n, n \geq 1\}$ be a sequence of closed sets such that $X = \underset{n}{\cup}F_n$.
Then at least one of the sets F_n has non empty interior.

Proof: Put $G_n = X\backslash F_n$, then we have $\underset{n}{\cap}G_n = \emptyset$. Since X is a Baire space, there is k such that $\overline{G_k} \neq X$; we deduce that $X\backslash\overline{G_k} \neq \emptyset$, but $X\backslash\overline{G_k} = X\backslash\overset{o}{G_k} = \overset{o}{F_k}.\blacksquare$

The prototype of Baire space is given by the following theorem:

Theorem 3.9.3. Every complete metric space (X,d) is a Baire space.

Proof: Let $\{G_n, n \geq 1\}$ be a sequence of dense open sets in X, we have to show that $\underset{n}{\cap}G_n$ is dense in X. Let V be non empty open set of X; since G_1 is dense, $V \cap G_1 \neq \emptyset$ and so there is an open ball B_1 with $\overline{B_1} \subset V \cap G_1$ and $\delta(\overline{B_1}) \leq 1$, where $\delta(\overline{B_1})$ is the diameter of $\overline{B_1}$.
Likewise, G_2 being dense, $\overline{B_1} \cap G_2 \neq \emptyset$ and there is an open ball B_2 with $\overline{B_2} \subset B_1 \cap G_2$ and $\delta(\overline{B_2}) \leq \dfrac{1}{2}$. We can iterate the process and get a sequence B_n of open balls satisfying:

$$\overline{B_n} \subset B_{n-1} \cap G_n \text{ and } \delta\left(\overline{B_n}\right) \le \frac{1}{n}.$$

The $\left(\overline{B_n}\right)$ is a decreasing sequence of closed sets such that $\delta\left(\overline{B_n}\right) \longrightarrow 0, n \longrightarrow \infty$. Since X is complete, this yields $\cap_n \overline{B_n} \ne \emptyset$ (Theorem **3.2.9**). But $\cap_n \overline{B_n} \subset V \cap \left(\cap_n G_n\right)$, so $V \cap \left(\cap_n G_n\right) \ne \emptyset$, since V is arbitrary we deduce the density of $\cap_n G_n$.∎

Examples 3.9.4. (a) The euclidean spaces $\mathbb{R}^n, n \ge 1$ are Baire spaces.
(b) The set $\mathbb{Q}$ of rationals is not a Baire space. Indeed we have $\mathbb{Q} = \cup_n \{r_n\}$ and each $\{r_n\}$ is closed with empty interior.
(b) In $\mathbb{R}$, the sets $\mathbb{R} \setminus \{x\}$ are open and dense, however $\cap_x \mathbb{R} \setminus \{x\} = \emptyset$, so countability is essential in Definition **3.9.1**.

Definition 3.9.5. A subset A of a topological space (X, τ) is said to be non dense if $\overset{o}{\overline{A}} = \emptyset$. The subset A is said to be of the first category if it can be written as a countable union of non dense sets.
A set is of the second category if its not of the first category.

Examples 3.9.6. (a) Every Baire space is of the second category
(Proposition **3.9.2**), in particular every complete metric space is of the second category (Theorem **3.9.3**).
(b) The set $\mathbb{Q}$ of rationals is of the first category.
(b) Every countable set of a Hausdorff non discrete space is of the first category.

Theorem 3.9.7. In a Baire space, every set of first category has empty interior.

Proof: Let A be of the first category in X then $A = \cup_n B_n$, where the B_n are non dense. Let U be an open set with $U \subset A$; we must prove that $U = \emptyset$. We have $U \subset \cup_n B_n \subset \cup_n \overline{B_n}$, so $\cap_n X \setminus \overline{B_n} \subset X \setminus U$. Since $\overset{o}{\overline{B_n}} = \emptyset$, the sets $X \setminus \overline{B_n}$ are dense and then $\cap_n X \setminus \overline{B_n}$ is also dense because X is a Baire space. We deduce that the closed set $X \setminus U$ is dense in X, which gives $X \setminus U = X$, that is $U = \emptyset$.∎

Remark 3.9.8. (a) Although a set A of the first category, in a Baire space, has empty interior, the closure $\overline{A}$ of A may have non empty interior. For example the set $\mathbb{Q}$ of rationals is of the first category in $\mathbb{R}$, however $\overset{o}{\overline{\mathbb{Q}}} = \mathbb{R}$.

(b) In a Baire space the complement of a set of the first category is a set of the second category.
(c) A set of the 2nd category may have empty interior as is shown by the set of irrationals in $\mathbb{R}$.

10. Exercises

54. Prove that every discrete space is a Baire space.
55. Let X be a Baire space and Y a topological space. Let $f : X \longrightarrow Y$ be a function, surjective, continuous and open. Prove that Y is a Baire space.
56. Prove that in a Baire space every countable intersection of dense open sets is of the 2nd category.
57. In a Baire space, give an example of a set of the 2nd category whose complement is not of the first category.

58. A perfect set E in a topological space is a closed set in which every point is a limit point. It is the same to say that a perfect set is a closed set which has no isolated points.

   (a) The set $\mathbb{Q}$ of rationals is not perfect although it has no isolated points.

   (b) If E has no isolated points prove that its closure $\overline{E}$ is a perfect set.

   (c) Every closed and bounded interval in $\mathbb{R}$ (and $\mathbb{R}$ itself!) is perfect.

   (d) A discrete space is not perfect.

CHAPTER 4

Compact Spaces and Locally Compact Spaces

The concept of compactness is an abstraction of the Heine-Borel property satisfied by any closed and bounded interval in $\mathbb{R}$ and formulated by:

If $[a,b] \subset \underset{\alpha}{\cup} V_\alpha$, where the V_α are open sets in $\mathbb{R}$, then there is a finite number $V_{\alpha_1}, ..., V_{\alpha_n}$ among the V_α such that $[a,b] \subset V_{\alpha_1} \cup ... \cup V_{\alpha_n}$. (see Theorem **4.7.3.** below)

A general statement of this property in a topological space leads to the concept of compactness which a considerable importance in analysis.

1. Compact Spaces

Definition 4.1.1. (a) A Hausdorff topological space (X, τ) is said to be compact if for any family $\{U_\alpha, \alpha \in A\}$ of open sets in X such that $X = \underset{\alpha \in A}{\cup} U_\alpha$, there is a finite subfamily $\{U_{\alpha_j}, 1 \leq j \leq n\}$ still with $X = \overset{n}{\underset{j=1}{\cup}} U_{\alpha_j}$. In other words, X is compact if every open cover $\{U_\alpha, \alpha \in A\}$ of X contains a finite subcover $\{U_{\alpha_j}, 1 \leq j \leq n\}$.

(b) A subset $K \subset X$ is compact if the subspace (K, τ_K), with the trace topology τ_K, is compact. Since every $V \in \tau_K$ is of the form $V = U \cap K$, $U \in \tau$, (b) is equivalent to:

$K \subset X$ is compact if for any family $\{U_\alpha, \alpha \in A\}$ of open sets in X such that $K \subset \underset{\alpha}{\cup} U_\alpha$, there is a finite subfamily $\{U_{\alpha_j}, 1 \leq j \leq n\}$ with $K \subset \overset{n}{\underset{j=1}{\cup}} U_{\alpha_j}$.

Definition 4.1.2. We say that a family of sets (E_α) has the finite intersection property if every finite subfamily of (E_α) has non empty intersection.

The following proposition is straightforward.

Proposition 4.1.3. In a Hausdorff space (X, τ) the following properties are equivalent:

(a) X is compact.

(b) Every family of closed sets with empty intersection contains a finite subfamily with empty intersection.

(c) Every family of closed sets having the finite intersection property has non empty intersection.

Examples 4.1.4. (a) Every finite set in a Hausdorff space is compact.

(b) A discrete space is compact if and only if it is finite.

(c) The space $\mathbb{R}$ is not compact, e.g the open covering $\{]-n, n[, n \geq 1\}$ has no finite subcover.

(d) A closed and bounded interval $[a,b]$ in $\mathbb{R}$ is compact by Heine-Borel theorem $[]$.

41

Theorem 4.1.5. Let X be a Hausdorff space and $K \subset X$ a compact set.
If $x \in X \backslash K$ there are open sets U, V in X such $x \in U, K \subset V$ and $U \cap V = \emptyset$.

Proof: Since X is Hausdorff, for each $y \in K$ there exist disjoint open sets U_y, V_y in X such that $x \in U_y, y \in V_y$. The open sets V_y cover K, and since K is compact, there is a finite number of them $V_{y_1}, ..., V_{y_n}$, with $K \subset V_{y_1} \cup ... \cup V_{y_n}$. Then the open sets $U = \cap U_{y_j}$ and $V = \cup V_{y_j}$ give the conclusion.∎

Corollary: In a Hausdorff space, every compact set is closed.

Proof: By the above theorem we have $x \notin K \implies x \notin \overline{K}$ so $\overline{K} \subset K$, that is $\overline{K} = K$, and K is closed.∎

Proposition 4.1.6. Let X be a compact space and $K \subset X$. Then the following properties are equivalent:
    (a) K is compact.
    (b) K is closed.

Proof: $(a) \implies (b)$ This is the above corollary.
Conversely if K is closed, let (G_α) be a covering of K by open sets of X. Then we have $X = \underset{\alpha}{\cup} G_\alpha \cup V$, with $V = X \backslash K$ open; since X is compact the family $\{(G_\alpha), V\}$ contains a finite subcover of X. If V is included in this subcover we can remove it and obtain a finite subcover of K, so K is compact.∎

Theorem 4.1.7. Every compact space is normal.

Proof: For normal space see Definition **2.5.7**, Chapter 2. Let F, H be closed sets in the compact space X, with $F \cap H = \emptyset$. By Proposition **4.1.6** F and H are compact; applying Theorem **4.1.5** we get, for each $x \in F$, open sets U_x, V_x in X such that $x \in U_x$, $H \subset V_x$ and $U_x \cap V_x = \emptyset$. Now the open covering (U_x) of F contains a finite subcovering $F \subset U_{x_1} \cup ... \cup U_{x_n}$. This gives open sets $U = \cup U_{x_i}$ and $V = \cap V_{x_i}$ satisfying $U \cap V = \emptyset$ and $F \subset U, H \subset V$.∎

This proof can easily be adapted to obtain:

Proposition 4.1.8. Let X be a Hausdorff space and C, K be compact sets in X such that $C \cap K = \emptyset$. Then there exist open sets U, V, with $U \cap V = \emptyset$ and $C \subset U, K \subset V$.∎

Theorem 4.1.9. Bolzano-Weierstrass
Let A be an infinite subset of a compact space X, then A has at least one limit point.

Proof: Suppose A has no limit point. Then for each $x \in X$ there is an open set V_x with $x \in V_x$, and $V_x \cap A$ contains at most one point. Then we have $X = \underset{x \in X}{\cup} V_x$; the compactness of X reduces the covering (V_x) to a finite one: $X = V_{x_1} \cup ... \cup V_{x_n}$, but then A would be equal to $(A \cap V_{x_1}) \cup ... \cup (A \cap V_{x_n})$ and would be finite contrary to the condition A infinite.∎

2. Exercises

59. Let (X, τ) be topological space and $K \subset A \subset X$. Prove that K is compact in (X, τ) if and only if K is compact in (A, τ_A).

60. Let F, K be subsets of the Hausdorff space X, such that F is closed and K compact. Prove that $F \cap K$ is compact.

61. In a Hausdorff space, prove that the union of a finite family of compact sets is a compact set.Give a simple example to show that this is not true in general for infinite families.

62. Let X be a Hausdorff space and let (K_α) be any family of compact sets in X.
(a) Prove that $\underset{\alpha}{\cap} K_\alpha$ is compact.

(b) If $\underset{\alpha}{\cap} K_\alpha = \emptyset$ then (K_α) contains a finite subfamily $G_{\alpha_1}, ..., G_{\alpha_n}$
such that $G_{\alpha_1} \cap ... \cap G_{\alpha_n} = \emptyset$.

(c) Deduce that for every sequence (K_n) of non empty compact sets such that $K_1 \supset K_2 \supset ...$, we have $\underset{n}{\cap} K_n \neq \emptyset$.

63. A Hausdorff space X is said to be regular if for any closed set F in X and any $x \notin F$, there exist open sets U, V such that $x \in U, F \subset V$ and $U \cap V = \emptyset$. This is an intermediate separation property between Hausdorff property and normality.
Let K be compact in a regular space X. Prove that for every open set U such that $K \subset U$, there is an open V satisfying $K \subset V \subset \overline{V} \subset U$. [$K \cap X \backslash U = \emptyset$, so apply regularity to each $x \in K$ with the closed set $X \backslash U$].

64. Let (X, τ) be a Hausdorff space. Let σ be a topology on X such that (X, σ) is compact. Prove that $\tau \subset \sigma \implies \tau = \sigma$ [every compact set for σ is compact for τ so every closed set for σ is closed for τ].

65. Let (I_n) be a sequence of compact intervals in $\mathbb{R}$ such that $I_n \supset I_{n+1}, \forall n$. Prove, without using Exercise **62**(c), that $\underset{n}{\cap} I_n \neq \emptyset$.
[Put $I_n = [a_n, b_n]$, the set $E = \{a_1, a_2, ...\}$ is bounded above by b_1. Then prove that $SupE \in \underset{n}{\cap} [a_n, b_n]$].

66. (The Cantor set). Let $X = [0, 1]$ be the unit interval of $\mathbb{R}$. Consider the sequence (E_n) of subsets of X defined as follows:

$$E_1 = X \backslash \left] \frac{1}{3}, \frac{2}{3} \right[= \left[0, \frac{1}{3} \right] \cup \left[\frac{2}{3}, 1 \right]$$

$$E_2 = E_1 \backslash \left] \frac{1}{9}, \frac{2}{9} \right[\cup \left] \frac{7}{9}, \frac{8}{9} \right[= \left[0, \frac{1}{9} \right] \cup \left[\frac{2}{9}, \frac{3}{9} \right] \cup \left[\frac{6}{9}, \frac{7}{9} \right] \cup \left[\frac{8}{9}, 1 \right]$$

and so on. We get a sequence (E_n) of subsets of X such that:

E_n is the union of 2^n disjoint closed intervals $I_{n,k}, k = 1, 2, ..., 2^n$, all having the same length $\dfrac{1}{3^n}$ and satisfying $E_1 \supset E_2 \supset ....$Now define the Cantor set by $C = \underset{n}{\cap} E_n$. Prove that

(a) E_n is compact for every n.
(b) C is not empty and compact.
(c) C has empty interior [C does not contain open interval].
(d) C is a perfect set (see Exercise **58**).

3. Compact Metric Spaces

Lemma 4.3.1. Let (X, d) be a compact metric space. For every $\epsilon > 0$ there is a finite covering for X by open balls with radius ϵ. In particular X is bounded.

Proof: The family of open balls $\{B(x, \epsilon), x \in X\}$ is an open covering for X which reduces to a finite one by compactness, thus we have $X = B(x_1, \epsilon) \cup ... \cup B(x_n, \epsilon)$, for some $x_1, ..., x_n$. Putting $M = \underset{i,j}{Max} \, d(x_i, x_j)$ we get $\delta(X) \leq M + 2\epsilon < \infty$.$\blacksquare$

Lemma 4.3.2. Let (V_α) be an open covering of a compact metric space (X, d). There exists $\lambda > 0$ such that $\forall x \in X \; \exists \alpha : B(x, \lambda) \subset V_\alpha$. We call λ the Lebesgue number associated to the open covering (V_α).

Proof: Since $X = \underset{\alpha}{\cup} V_\alpha$, we have for each $x \in X$ there is α, and there is $r_x > 0$ such that $B(x, r_x) \subset V_\alpha$. From the open covering $\left\{ B\left(x, \dfrac{r_x}{2}\right), x \in X \right\}$ we can extract a finite subcovering, say, $\left\{ B\left(x_i, \dfrac{r_{x_i}}{2}\right), 1 \leq i \leq n \right\}$. Define λ by $\lambda = Min\left\{\dfrac{r_{x_i}}{2}, 1 \leq i \leq n\right\}$. We prove that such λ works: indeed if $x \in X$ there is i with $x \in B\left(x_i, \dfrac{r_{x_i}}{2}\right)$, so that for $z \in B(x, \lambda)$ we will have
$$d(z, x_i) \leq d(z, x) + d(x, x_i) \leq \lambda + \frac{r_{x_i}}{2} \leq r_{x_i}.$$
Consequently $B(x, \lambda) \subset B(x_i, r_{x_i}) \subset V_\alpha$, for some α.$\blacksquare$

Theorem 4.3.3. Every compact metric space (X, d) is separable. In particular such space is second countable.

Proof: From Lemma **4.2.1** for each n the space X has a finite covering by open balls with radius $\dfrac{1}{n}$. Let C_n be the set of centers of these balls; then the set $D = \underset{n}{\cup} C_n$ is countable and dense in X; indeed if $B(x, \alpha)$ is any open ball, there is $n \geq 1$ such that $\dfrac{1}{n} \leq \alpha$, and, from the definition of C_n, there is $x_{m_n} \in C_n$ with $d(x, x_{m_n}) < \dfrac{1}{n} \leq \alpha$, so that $x_{m_n} \in B(x, \alpha)$, consequently X is separable. By Theorem **3.7.3** Chapter **3**, X is second countable.$\blacksquare$

Theorem 4.3.4. Let (X, d) be a metric space. The following are equivalent:
    (a) X is compact.
    (b) Every infinite subset $A \subset X$ has at least one limit point.

[Note that $(a) \implies (b)$ is true in every compact space by Theorem **4.1.9**.]

Proof: see $[\,]$.$\blacksquare$

Corollary: Let (X, d) be a metric space. The following are equivalent:
    (a) X is compact.
    (b) Every infinite sequence in X contains a convergent subsequence.

Definition 4.3.5. A metric space (X, d) is totally bounded, or precompact, if for every $\epsilon > 0$, X has a finite covering by open balls with radius ϵ.

Theorem 4.3.6. Let (X, d) be a metric space. The following are equivalent:
    (a) X is compact.
    (b) X is complete and totally bounded.

Proof: If X is compact then it is totally bounded, by Lemma **4.3.1**. On the other hand let (x_n) be a Cauchy sequence in X, then from the corollary of Theorem **4.3.4**, (x_n) contains a convergent subsequence (x_{n_k}), say with limit x. Then we have $d(x_n, x) \leq d(x_n, x_{n_k}) + d(x_{n_k}, x) \longrightarrow 0, k \longrightarrow \infty, n \longrightarrow \infty$, so X is complete.

To prove the converse we also use the above corollary, that is we show that every infinite sequence (x_n) in X contains a convergent subsequence. Since X is totally bounded, for each $k \geq 1$, there is a covering $\mathcal{B}_k$ for X with open balls of radius $\dfrac{1}{k}$. Consequently there is $B_1 \in \mathcal{B}_1$ such that $B_1 \cap \{x_n, n \geq 1\}$ is infinite. Likewise there is $B_2 \in \mathcal{B}_2$ such that $B_2 \cap B_1 \cap \{x_n, n \geq 1\}$ is infinite, and so on. We get a sequence (B_n) of open balls such that for every p, $B_1 \cap B_2 \cap ... \cap B_p \cap \{x_n, n \geq 1\}$ is infinite. At this stage define the subsequence $(x_{n_k}) \subset (x_n)$ by:

choose x_{n_1} in $B_1 \cap \{x_n, n \geq 1\}$

since $B_1 \cap B_2 \cap \{x_n, n \geq 1\}$ is infinite, there is $n_2 > n_1$ with $x_{n_2} \in B_1 \cap B_2$,

since $B_1 \cap B_2 \cap B_3 \cap \{x_n, n \geq 1\}$ is infinite, there is $n_3 > n_2$ with $x_{n_3} \in B_1 \cap B_2 \cap B_3$.

In this way we get a subsequence (x_{n_k}) of (x_n) such that $1 \leq n_1 < n_2 < n_3 < ...$ and $x_{n_k} \in B_1 \cap B_2 \cap ... \cap B_k$ for every k.

It is clear that for $n_i \geq n_j$ we have $x_{n_i}, x_{n_j} \in B_j$ and then

$$d\left(x_{n_i}, x_{n_j}\right) < \frac{2}{j} \longrightarrow 0, n_i, n_j \longrightarrow \infty, \text{ so } (x_{n_k}) \text{ is Cauchy. We deduce that } x_{n_k}$$

converges since X is complete.∎

4. Exercises

67. Let (X, d) be a metric space and let (x_n) be a convergent sequence in X with limit x. Prove that the set $E = \{x_n, n \geq 1\} \cup \{x\}$ is compact.

68. Prove that in a totally bounded space every infinite sequence contains a Cauchy subsequence.

69. We say that a topological space is countably compact if every countable open covering contains a finite subcovering.

(a) Prove that X is countably compact if and only if every infinite sequence has a limit point.

(b) Prove that in a metric space the compactness is equivalent to the countable compactness.

5. Continuous Functions on Compact Spaces

Theorem 4.5.1. Let be a continuous function from a compact space X into a Hausdorff space Y. Then we have:

    (a) The image $f(X)$ of X is compact.

    (b) The function f is closed, that is the image by f of a closed set is closed.

    (c) If f is a bijection then f is an homeomorphism from X onto Y.

Proof: (a) Let (U_α) be an open covering of $f(X)$. Since f is continuous, the family $\left(f^{-1}(U_\alpha)\right)$ is an open covering of X which can be reducsd to a finite one say $f^{-1}(U_{\alpha_1}), ..., f^{-1}(U_{\alpha_n})$ by the compactness of X; then it is clear that $U_{\alpha_1}, ..., U_{\alpha_n}$ is a finite covering of $f(X)$.

(b) Let $A \subset X$ be closed then A is compact (Proposition **4.1.6**) and $f(A)$ is compact by (a); since Y is Hausdorff $f(A)$ is closed by corollary of Theorem **4.1.5**.

(c) If f is bijective, let $A \subset X$ be closed then $\left(f^{-1}\right)^{-1}(A) = f(A)$ is closed in Y by (b), so f^{-1} is continuous.∎

Theorem 4.5.2. Every continuous function $f : X \longrightarrow \mathbb{R}$ from a compact metric space into $\mathbb{R}$ is bounded an achieves its bounds.

Proof: By part (a) of the above Theorem $f(X)$ is a compact subset of $\mathbb{R}$, and so, by Heine-Borel Theorem $[]$, $f(X)$ is a closed and bounded interval. Put $M = \sup_x f(x)$, $m = \inf_x f(x)$; we prove that M is attained, that is, there is $a \in X$ with $f(a) = M$; the proof for m is similar. The nature of M gives a sequence $(x_n) \subset X$ such that $\lim_n f(x_n) = M$. From the corollary of Theorm **4.3.4**, we get a convergent subsequence $(x_{n_k}) \subset (x_n)$ with $\lim_n x_{n_k} = a$, and the continuity of f implies $\lim_k f(x_{n_k}) = f(a)$ so $f(a) = M$ since $\lim_n f(x_n) = M$.$\blacksquare$

Theorem 4.5.3. Every continuous function $f : X \longrightarrow Y$ from a compact metric space (X, d) into a metric space (Y, δ) is uniformly continuous.

Proof: We have to show that:
$$\forall \epsilon > 0, \exists \lambda = \lambda(\epsilon) > 0 : d(x, x') < \lambda \implies \delta(f(x), f(x')) < \epsilon$$
Let $\left\{ B\left(y, \frac{\epsilon}{2}\right), y \in Y \right\}$ be the open covering of Y by the open balls $B\left(y, \frac{\epsilon}{2}\right)$. Let $\lambda = \lambda(\epsilon)$ be the Lebesgue number associated to the open covering $\left\{ f^{-1}\left(B\left(y, \frac{\epsilon}{2}\right)\right), y \in Y \right\}$ of X (see Lemma **4.3.2** for the definition of the Lebesgue number). Then, by Lemma **4.3.2**, every open ball $B(x, \lambda)$ of X is contained in at least one set $f^{-1}\left(B\left(y, \frac{\epsilon}{2}\right)\right)$. Consequently if $d(x, x') < \lambda$ we will have $f(x), f(x') \in B\left(y, \frac{\epsilon}{2}\right)$ for some $y \in Y$; this yields
$$\delta(f(x), f(x')) \le \delta(f(x), y) + \delta(y, f(x')) < \frac{\epsilon}{2} + \frac{\epsilon}{2} = \epsilon.\blacksquare$$

Theorem 4.5.4. Let X be a compact topological space and let (f_n) be a sequence of continuous functions from X into a metric space (Y, d) satisfying the following condition:
(Equicontinuity):
  For each $x \in X$ and each $\epsilon > 0$ there is an open set U with
  $x \in U$ and $d(f_n(x), f_n(y)) < \epsilon, \forall y \in U, \forall n \ge 1$
Suppose that f_n converges pointwise to some function $f : X \longrightarrow Y$, then f is continuous and the convergence of f_n to f is uniform on X.

Proof: Fix $\epsilon > 0$. By the equicontinuity, for each $x \in X$ there is an open set U_x with $x \in U_x$ and $d(f_n(x), f_n(y)) < \frac{\epsilon}{3}, \forall y \in U_x, \forall n \ge 1$. Since $f_n(x) \longrightarrow f(x)$ and $f_n(y) \longrightarrow f(y)$ as $n \longrightarrow \infty$, we deduce that $d(f(x), f(y)) < \frac{\epsilon}{3}, \forall y \in U_x$. This gives the continuity of f. Next since X is compact the open covering $\{U_x, x \in X\}$ can be reduced to a finite one, say, $\{U_{x_1}, ..., U_{x_{n_0}}\}$. Since $f_n(x) \longrightarrow f(x)$ on X, there is $N = N_\epsilon \ge 1$ such that $\forall n \ge N, d(f_n(x_i), f(x_i)) < \frac{\epsilon}{3}, i = 1, ..., n_0$. On the other hand $\forall y \in X$ there is $1 \le i \le n_0$ such that $y \in U_{x_i}$ then we have
$\forall n \ge N, d(f_n(y), f(y)) \le$
  $d(f_n(y), f_n(x_i)) + d(f_n(x_i), f(x_i)) + d(f(x_i), f(y)) < \epsilon$, this proves the uniform convergence of f_n to f.$\blacksquare$

6. Exercises

70. Let X be a compact topological space and let (f_n) be a sequence of continuous functions from X into $\mathbb{R}$ such that $f_n \leq f_{n+1}$ on $X, \forall n \geq 1$. Suppose there is a continuous $g : X \longrightarrow \mathbb{R}$ such that $\lim_n f_n(x) = g(x)$, $\forall x \in X$.

Prove that the convergence of f_n to g is uniform. [put $\varphi_n = g - f_n$, then $\varphi_n \geq \varphi_{n+1}$ on $X, \forall n \geq 1$ and $\varphi_n(x) \longrightarrow 0$, $\forall x \in X$. Prove that for each $\epsilon > 0$ we have $X = \underset{n}{\cup} \{x : \varphi_n(x) < \epsilon\}$. Since X is compact conclude that there is $N = N_\epsilon \geq 1$ such that $\forall n \geq N$, $X = \{x : \varphi_n(x) < \epsilon\}$.]

71. Let (X, d) be a compact metric space and let $f : X \longrightarrow X$ be a function satisfying $d(f(x), f(y)) = d(x, y)$, $\forall x, y$. Prove that f is an homeomorphism and that $d\left(f^{-1}(x), f^{-1}(y)\right) = d(x, y)$, $\forall x, y$. [To prove that f is onto, suppose there is $z \notin f(X)$, this gives $d\left(f^{n-1}(z), f^n(z)\right) = d(z, f(z))$, $\forall n \geq 1$. Since X is compact metric the sequence $f^n(z)$ contains a convergent subsequence (corollary of Theorem **4.3.4.**) and we get $d(z, f(z)) = 0$, whence a contradiction with the fact $z \notin f(X)$].

72. Let X be a topological space. We say that a function $f : X \longrightarrow X$ is a local homeomorphism at $x \in X$ if there is an open set U with $x \in U$ such that f is an homeomorphism from U onto $f(U)$. f is said to be a local homeomorphism on X if it is a local homeomorphism at every $x \in X$.
(a) Prove that if X is compact and $f : X \longrightarrow X$ is a local homeomorphism, then for every $x \in X$ the fibre $f^{-1}(x)$ is finite.
(b) Let (X, d) be a compact metric space and let $f : X \longrightarrow X$ be a local homeomorphism, prove that there is $\alpha > 0$ such that:
$$\forall x, y \in X, \ f(x) = f(y) \implies d(x, y) > \alpha$$
[use contradiction reasoning in both $(a), (b)$].

73. Let $\varphi : X \longrightarrow X$ be a continuous surjection of a compact metric space (X, d). We say that φ is expansive with constant $\alpha > 0$ if:
$$\forall x \neq y \ \exists n = n(x, y) \geq 1 : d(\varphi^n(x), \varphi^n(y)) > \alpha$$
If φ is expansive, prove that for every $x \in X$ the fibre $f^{-1}(x)$ is finite.

74. Let X be a compact space and let $x, y \in X$. Suppose that $f(x) = f(y)$ for every bounded continuous function $f : X \longrightarrow \mathbb{R}$. Prove that $x = y$. [use Urysohn Lemma **2.5.10** Chapter 2].

7. Product of Compact Spaces

Let $\{X_\alpha, \tau_\alpha, \alpha \in A\}$ be a family of topological spaces and form the cartesian product $X = \Pi X_\alpha$ of the sets X_α. Let $\pi_\alpha : X \longrightarrow X_\alpha$ be the $\alpha-$coordinate projection. We equip $X = \underset{\alpha}{\Pi} X_\alpha$ with the product topology $\underset{\alpha}{\otimes} \tau_\alpha$, making all the projections π_α continuous. Let us recall (Remark **2.3.14.** Chapter **2**) that the open elementary sets $[B_{\alpha_1}, B_{\alpha_2}, ..., B_{\alpha_n}] = \overset{n}{\underset{1}{\cap}} \pi_{\alpha_i}^{-1}(B_{\alpha_i})$ form a base of $\underset{\alpha}{\otimes} \tau_\alpha$.

Theorem 4.7.1. (Alexander)
Let X be a Hausdorff space and let $\mathcal{F}$ be a subbase for the topology of X. Suppose that any covering of X by open sets in $\mathcal{F}$ contains a finite subcovering. Then X is compact.
Proof: See [] .■

Theorem 4.7.2. (Tychonoff)

The product $\left(\Pi_\alpha X_\alpha, \otimes_\alpha \tau_\alpha\right)$ of every family $\{X_\alpha, \tau_\alpha, \alpha \in I\}$ of compact spaces is compact.

Proof: Note first that $\left(\Pi_\alpha X_\alpha, \otimes_\alpha \tau_\alpha\right)$ is Hausdorff (Theorem **2.5.5.** (b) Chapter **2**). On the other hand, Alexander Theorem allows one to consider open coverings for $\Pi_\alpha X_\alpha$ by open sets in the family $\mathcal{F} = \left\{\pi_\alpha^{-1}(U_\alpha), U_\alpha \in \tau_\alpha, \alpha \in I\right\}$ which is a subbase for the product topology $\otimes_\alpha \tau_\alpha$. So let $\mathcal{U}$ be an open covering for $\Pi_\alpha X_\alpha$ by open sets in $\mathcal{F}$ and for each α put $\mathcal{U}_\alpha = \left\{U \in \tau_\alpha : \pi_\alpha^{-1}(U) \in \mathcal{U}\right\}$; then there is $\alpha \in I$ such that $\mathcal{U}_\alpha$ constitutes an open covering for X_α (why?). Since X_α is compact there is $\{U_1, ..., U_n\} \subset \mathcal{U}$ with $X_\alpha = \overset{n}{\underset{1}{\cup}} U_i$, then it is clear that $\left\{\pi_\alpha^{-1}(U_1), ..., \pi_\alpha^{-1}(U_n)\right\}$ is a finite subcovering of $\mathcal{U}$ for $\Pi_\alpha X_\alpha$.∎

Theorem 4.7.3. (Heine-Borel)

A subset K in $\mathbb{R}^n$ is compact if and only if it closed and bounded.

Proof: If K is compact, by Lemma **4.3.1**, K is bounded and K is closed by the corollary of Theorem **4.1.5.** Conversely suppose K closed and bounded. We have K bounded $\implies K \subset \overset{n}{\underset{1}{\Pi}} [a_i, b_i] = A$; since K is closed in $\mathbb{R}^n$, it is closed in A which is compact, so K is compact by Proposition **4.1.6.**∎

8. Exercises

75. Prove that the product set $[0, 1]^{\mathbb{N}}$ is compact.

76. Prove that in the space $\mathbb{R}^n$ every bounded infinite sequence contains a convergent subsequence.

9. Locally Compact Spaces

Definition 4.9.1. We say that a subset A of a topological space X is relatively compact if the closure $\overline{A}$ is compact.

Definition 4.9.2. A topological space X is locally compact if each of its points has an open neighborhood relatively compact.

Examples 4.9.3. (a) Every compact space is locally compact.
(b) $\mathbb{R}^n$ is locally compact.
(b) Every infinite discrete space is locally compact but not compact.

Theorem 4.9.4. Let X be a locally compact space. Then for every $x \in X$ and every open neighborhood U of x, there is an open set V with $\overline{V}$ compact and such that $x \in V \subset \overline{V} \subset U$.

Proof: Let W be a relatively compact open neighborhood of x and put $G = U \cap W$; then G is an open neighborhood of x contained in U and satisfying $\overline{G} \subset \overline{W}$; since $\overline{W}$ is compact and $\overline{G}$ closed, we deduce that $\overline{G}$ is compact. Consider the boundary of G that is $\Gamma = \overline{G} \cap \overline{X \backslash G}$, we have $\Gamma \subset \overline{G}$ and Γ closed, so Γ is compact. If $\Gamma = \emptyset$ we have $\overline{G} = G \cup \Gamma = G$, and the open set $V = G$ works. If $\Gamma \neq \emptyset$ then for each $y \in \Gamma$ we have $y \neq x$, and since X is Hausdorff, there are open neighborhoods

V_y, H_y of x and y respectively such that $V_y \cap H_y = \emptyset$. We can assume $V_y \subset G$, otherwise we take $V_y \cap G$ instead of V_y. Now the open covering $\{H_y, y \in \Gamma\}$ of Γ can be reduced to a finite one by the compactness of Γ, say, $\Gamma \subset H_{y_1} \cup ... \cup H_{y_n} = H$. Put $V = V_{y_1} \cap ... \cap V_{y_n}$, then $V \subset G$ and $\overline{V} \subset \overline{G}$, so $\overline{V}$ is compact. On the other hand $V \cap H = \emptyset$, this yields $V \subset X \backslash H$ and since $X \backslash H$ is closed, $\overline{V} \subset X \backslash H$. We deduce that $\overline{V} \subset \overline{G} \cap X \backslash H \subset \overline{G} \cap X \backslash \Gamma = G$, where the equality comes from this one $\overline{G} = G \cup \Gamma$, so the open set $V = V_{y_1} \cap ... \cap V_{y_n}$ gives the conclusion.∎

If we take in the above theorem a compact set in X instead of a point x, we get:

Theorem 4.9.5. Let X be a locally compact space, and $K \subset X$ compact. Then, if U is open with $K \subset U$, there is an open set V such that
$\overline{V}$ is compact and $K \subset V \subset \overline{V} \subset U$.

Proof: By Theorem **4.9.4.**, for each $x \in K$ there is an open set V_x with $\overline{V}_x$ compact and such that $x \in V_x \subset \overline{V}_x \subset U$. As K is compact the open covering $\{V_x, x \in K\}$ of K can be reduced to a finite one: $K \subset V_{x_1} \cup ... \cup V_{x_n} = V \subset U$. The open set $V = V_{x_1} \cup ... \cup V_{x_n}$ satisfies the conclusion since $\overline{V} = \overline{V_{x_1}} \cup ... \cup \overline{V_{x_n}}$ is compact.∎

Corollary 1. A locally compact space has a base of relatively compact open sets.

Proof: Let $\mathcal{B}$ be the family of all relatively compact open sets of X. By Theorem **4.9.4**, for every open set U and every $x \in U$, there is $V \in \mathcal{B}$ such that $x \in V \subset U$, this proves that $\mathcal{B}$ is a base.∎

Corollary 2. Let X be a locally compact space with a countable base. Then X has a countable base of relatively compact open sets.

Proof: Let $\mathcal{U} = \{U_n, n \geq 1\}$ a countable base for X. By Theorem **4.9.4**, for each U_n and each $y \in V_y$ there is an open set V_y with $\overline{V}_y$ compact and $y \in V_y \subset \overline{V}_y \subset U_n$. But U_n considered as a subspace of X, has itself a countable base, so the covering $\{V_y, y \in U_n\}$ of U_n contains a countable subcovering $\{V_{n,m}, m \geq 1\}$. Then it is clear that the family of relatively compact open sets $\{V_{n,m}, n \geq 1, m \geq 1\}$ is a countable base for X.∎

The existence of non constant continuous functions on a topological space is not a trivial problem in general. For normal spaces, in particular for compact spaces, examples of such functions are given by Urysohn lemma (Theorem **2.5.10** Chapter **2**). We give below an adaptation of Urysohn lemma to a locally compact space (for an example of a locally compact space which is not normal see []). Let us start with the following preliminary:

Proposition 4.9.6. Let C be a closed set in a topological space X and $f : C \longrightarrow \mathbb{R}$ a continuous function on C such that $f(x) = 0$ for $x \in \partial C$. Define $F : X \longrightarrow \mathbb{R}$ by $F : X \longrightarrow \mathbb{R}$ by

$$
\begin{aligned}
F(x) &= f(x), x \in C \\
F(x) &= 0, x \in X \backslash C
\end{aligned}
$$

Then F is a continuous extension of f to X.

Theorem 4.9.7. Let X be a locally compact space, and $K \subset X$ compact, U open with $K \subset U$. There is a continuous function $F : X \longrightarrow \mathbb{R}$ such that:
$0 \leq F(x) \leq 1, \forall x \in X \quad F(x) = 1, \forall x \in K \quad F(x) = 0, \forall x \in X \backslash U$

Proof: By Theorem **4.9.5.** there is an open set V such that $\overline{V}$ is compact and $K \subset V \subset \overline{V} \subset U$. Put $\overline{V} = C$ and $\Gamma = \partial C$. Since C is compact it is normal; on the other hand Γ and K are disjoint closed subsets of C, so by Urysohn lemma there is a continuous function $f : C \longrightarrow [0,1]$ on C such that
$$f(x) = 1, \forall x \in K \text{ and } f(x) = 0, \forall x \in \Gamma$$
Let us extend f to the function $F : X \longrightarrow [0,1]$ given by $F(x) = f(x), x \in C$ and $F(x) = 0, x \in X \backslash C$. The function $F : X \longrightarrow [0,1]$ so defined meets the conclusion by Proposition **4.9.6.**∎

Remark 4.9.8. The support of the function F is the closed set $\operatorname{supp}(F) = \overline{\{x \in X : F(x) \neq 0\}}$. The above construction shows that $\operatorname{supp}(F)$ is compact and satisfies $\operatorname{supp}(F) \subset C \subset U$.

Theorem 4.9.9. Every locally compact space is a Baire space.

Proof: From the definition of a Baire space, we have to show that for every sequence (D_n) of dense open sets in X, the intersection $\bigcap_n D_n$ is dense. Let U be any non empty open set. We have $U \cap D_1 \neq \emptyset$, since D_1 is dense; by Theorem **4.9.4**, there a relatively compact open set V_1 such that $\overline{V_1} \subset U \cap D_1$. Similarly, using V_1 and D_2, there is a relatively compact open set V_2 such that $\overline{V_2} \subset V_1 \cap D_2$. Continuing the process, we get a sequence (V_n) of relatively compact open sets with $\overline{V_n} \subset V_{n-1} \cap D_n, n \geq 1$ and $V_0 = U$. Since the $\overline{V_n}$ are closed in the compact set $\overline{V_1}$, we have $\bigcap_n \overline{V_n} \neq \emptyset$ (why?). Since $\bigcap_n \overline{V_n} \subset U \cap \bigcap_n D_n \neq \emptyset$.∎

10. Exercises

77. Let X be a locally compact space and $K \subset X$ compact. Prove that there exist an open set U, with $\overline{U}$ compact and $K \subset U$. [for each $x \in K$ there is a relatively compact open set U_x containing x. Then consider a finite covering of K by the open sets U_x]

78. Let U be an open set in a locally compact space. Prove that the subspace U endowed with the trace topology is locally compact.

79. Let X be a locally compact space and $K \subset X$ compact. Let $U_1, U_2, ..., U_n$ be open sets such that $K \subset U_1 \cup U_2 \cup ... \cup U_n$. Prove that for each $i = 1, 2, ..., n$, there is an open set W_i satisfying: $\overline{W_i}$ compact, $\overline{W_i} \subset U_i$, and $K \subset W_1 \cup W_2 \cup ... \cup W_n$. [use Theorem **4.9.4**].

80. Consider the data of Exercise **79**. Prove that there exist n continuous functions $f_1, f_2, ..., f_n : X \longrightarrow [0,1]$ such that:
$$\operatorname{supp}(f_i) \subset U_i, 1 \leq i \leq n \text{ and } f_1(x) + f_2(x) + ... + f_n(x) = 1, \forall x \in K.$$
Such family of functions $\{f_1, f_2, ..., f_n\}$ is called partition of unity.
[use Theorem **4.9.7** and Remark **4.9.8**: for each $i = 1, 2, ..., n$, there is a continuous function $g_i : X \longrightarrow [0,1]$ with $g_i = 1$ on $\overline{W_i}$ and $\operatorname{supp}(g_i) \subset U_i$, then define $f_1 = g_1, f_2 = (1 - g_1) g_2, ..., f_n = (1 - g_1)(1 - g_2)...(1 - g_{n-1}) g_n$].

81. Let X, Y be locally compact spaces. Prove that the product $X \times Y$ endowed with the product topology is locally compact.

11. Compactification

Definition 4.11.1. Let X be a topological space. A compactification of X is a pair $\left(\widetilde{X}, \varphi\right)$, where $\widetilde{X}$ is a compact space and φ is an homeomorphism from X onto a dense subspace of $\widetilde{X}$.

Theorem 4.11.2. (Alexandroff compactification)

(1) (Existence) Every locally compact space has a compactification $\left(\widetilde{X}, \varphi\right)$ such that $\widetilde{X} \backslash X$ is a singleton.

(2) (Uniqueness) If $\left(\widetilde{X}, \varphi\right), \left(\widetilde{Y}, \psi\right)$ are two compactifications of X, there is an homeomorphism f from $\widetilde{X}$ onto $\widetilde{Y}$ such that the restriction of f to X is the identity map.

Usually the point $\widetilde{X} \backslash X$ is denoted by ∞, so we have $\widetilde{X} = X \cup \infty$.

Proof: See[] .

Example 4.11.3. Let S^n be the unit sphere in $\mathbb{R}^{n+1}$:
$$S^n = \left\{ (x_1, x_2, ..., x_{n+1}) \in \mathbb{R}^{n+1} : x_1^2 + x_2^2 ... + x_{n+1}^2 = 1 \right\}$$
Consider the projection $p : S^n \backslash (0, 0, ..., 1) \longrightarrow \mathbb{R}^n$ given by:
$$p(x_1, x_2, ..., x_{n+1}) = \left(\frac{x_1}{1 - x_{n+1}}, \frac{x_2}{1 - x_{n+1}}, ..., \frac{x_n}{1 - x_{n+1}} \right)$$
A straightforward checking shows that:

p is an homeomorphism

S^n is compact (closed and bounded in $\mathbb{R}^{n+1}$)

$S^n \backslash (0, 0, ..., 1)$ is dense in S^n

Then by Definition **4.11.1** and the uniqueness part of the above theorem, we deduce that S^n is the Alexandroff compactification of $\mathbb{R}^n$.

CHAPTER 5

Banach Spaces

1. Normed spaces

In what follows X is a vector space on the scalar field $\mathbb{R}$ or $\mathbb{C}$, whose nul vector will be denoted by 0. The notion of norm allows to define on X a metrizable topology compatible with the vector space structure.

Definition 5.1.1. A norm on the vector space X is a function $x \longrightarrow \|x\|$ from X into $\mathbb{R}$ satifying the conditions:

(a) $\|x\| \geq 0, \forall x$ and $\|x\| = 0 \iff x = 0$.

(b) $\|\lambda \cdot x\| = |\lambda| \cdot \|x\|, \forall x \in X$ and every scalar λ.

(c) $\|x + y\| \leq \|x\| + \|y\|, \forall x, y \in X$ (subadditivity property).

A vector space X equipped with a norm $\|\cdot\|$ is called
normed space and will be denoted by $X, \|\cdot\|$.

Let us point out right now that any norm makes the set X a metric space:

Proposition 5.1.2. Let $X, \|\cdot\|$ be a normed space. Define the function
$d : X \times X \longrightarrow \mathbb{R}$ by $d(x, y) = \|x - y\|$ then:

(a) d is a metric on X

(b) The function $x \longrightarrow \|x\|$ is uniformly continuous on the metric space X, d.

Proof: (a) is a consequence of properties $(a), (b), (c)$ of a norm.

(b) By the subadditivity property of the norm, we have $\forall x, y \in X$:

$\|x\| = \|y + x - y\| \leq \|y\| + \|x - y\|$

$\|y\| = \|x + y - x\| \leq \|x\| + \|y - x\|$

so we deduce that $|\|x\| - \|y\|| \leq \|x - y\| = d(x, y)$, and the uniform continuity of the function $x \longrightarrow \|x\|$ follows.$\blacksquare$

Example 5.1.3. (1) For any real $1 \leq p < \infty$, define $\|x\|_p$ on $\mathbb{R}^n$ by:

$$\|x\|_p = \left(\sum_{i=1}^{n} |x_i|^p \right)^{\frac{1}{p}} \qquad x = (x_1, x_2, ..., x_n) \in \mathbb{R}^n.$$

Then $\|x\|_p$ is a norm on $\mathbb{R}^n$. It is immediate for $p = 1$. If $p > 1$ let $q > 1$ be the conjugate of p, that is q satisfies the relation $\dfrac{1}{p} + \dfrac{1}{q} = 1$.

Then we have $\forall u, v \geq 0$, $u \cdot v \leq \dfrac{u^p}{p} + \dfrac{v^q}{q}$ and this relation gives the so called Holder inequality:

$$x, y \in \mathbb{R}^n, \quad \sum_{i=1}^{n} |x_i \cdot y_i| \leq \|x\|_p \cdot \|y\|_q$$

which in turn gives Minkowski inequality:

$$x, y \in \mathbb{R}^n, \quad \|x + y\|_p \leq \|x\|_p + \|y\|_p$$

that is, the subadditivity of $x \longrightarrow \|x\|_p$.

These are part of general convex inequalities [] .

For $p = 2$ we get the euclidean norm of $\mathbb{R}^n$ $\|x\|_2 = \left(\sum_{i=1}^{n} |x_i|^2 \right)^{\frac{1}{2}}$.

On the other hand, for $p = \infty$, we define $\|x\|_\infty$ by $\|x\|_\infty = \underset{1 \leq i \leq n}{Max} |x_i|$. It not difficult to check that $\|x\|_\infty$ is a norm on $\mathbb{R}^n$ satisfying:

$$\frac{1}{\sqrt{n}} \|x\|_\infty \leq \frac{1}{\sqrt{n}} \|x\|_2 \leq \|x\|_\infty$$

$$\|x\|_\infty \leq \|x\|_1 \leq n \|x\|_\infty$$

(2) l_p spaces, $1 \leq p < \infty$

Consider the set of real sequences given by $l_p = \left\{ (x_n) \in \mathbb{R}^{\mathbb{N}} : \sum_n |x_n|^p < \infty \right\}$.

If $x = (x_n)$, $y = (y_n)$ are in l_p and $\lambda \in \mathbb{R}$, put $x + y = (x_n + y_n)$ and $\lambda x = (\lambda x_n)$. Then by the convex inequality, $(u + v)^p \leq 2^{p-1} (u^p + v^p)$, valid for $u, v \geq 0$, we deduce that the above operations make l_p a vector space with nul vector $0 = (0, 0...)$.

For $x = (x_n) \in l_p$ we put $\|x\|_p = \left(\sum_n |x_n|^p \right)^{\frac{1}{p}}$. Then it is not difficult to prove that $\|x\|_p$ is a norm on l_p; to see the subadditivity property, consider the inequality $\|x + y\|_p \leq \|x\|_p + \|y\|_p$ valid for $x, y \in \mathbb{R}^n$ and for all $n \geq 1$, (example (1)), and then make n goes to ∞.

(3) l_∞ space:

Let us define the space l_∞ by the following ingredients:

$l_\infty = \left\{ (x_n) \in \mathbb{R}^{\mathbb{N}} : \underset{n}{Sup} |x_n| < \infty \right\}$, that is the set of bounded real sequences. If $x = (x_n)$, $y = (y_n)$ are in l_∞ and $\lambda \in \mathbb{R}$, put $x + y = (x_n + y_n)$, $\lambda x = (\lambda x_n)$, and $\|x\|_\infty = \underset{n}{Sup} |x_n|$. It is straightforward that this data makes l_∞ a normed space with $l_p \subset l_\infty$, for $1 \leq p < \infty$.

(4) Let X be a topological space and consider the set $C_b(X)$ of all continuous and bounded functions $f : X \longrightarrow \mathbb{R}$. If $f, g \in C_b(x)$ and $\lambda \in \mathbb{R}$, define $f + g$, and λf in the usual way. Next put $\|f\|_\infty = \underset{x \in X}{Sup} |f(x)|$, then $C_b(x)$ is a vector space and $\|f\|_\infty$ is a norm on $C_b(x)$. Let us point out the following simple fact:

If $f_n, f \in C_b(X)$, then we have $\|f_n - f\|_\infty \longrightarrow 0, n \longrightarrow 0$ if and only if f_n converges uniformly to f on X.

(5) Let $C\left[0,1\right]$ be the space of all continuous functions $f : \left[0,1\right] \longrightarrow \mathbb{R}$, and define $\|f\|_1 = \int_0^1 f\left(x\right)dx$ for $f \in C\left[0,1\right]$ (Riemann integral). Then $\|f\|_1$ is a norm on $C\left[0,1\right]$ and we have $\|f\|_1 \leq \|f\|_\infty$.

Notation: Let E, F be subsets of a vector space X and $\Lambda \subset \mathbb{R}$ or $\mathbb{C}$, in the sequel we use the following notations:

$E + F = \{a + b : a \in E, b \in F\}$

$\Lambda \cdot E = \{\lambda a : a \in E, \lambda \in \Lambda\}$

Definition 5.1.4. A subset E of a vector space X is said to be convex if for every $x, y \in E$ and $0 \leq \lambda \leq 1$, $\lambda x + \left(1 - \lambda\right)y \in E$; in other words E is convex if $\lambda E + \left(1 - \lambda\right)E \subset E$ for all $0 \leq \lambda \leq 1$.

Example 5.1.5. (a) Every vector subspace of X is convex.
(b) If E is convex then $a + E$ is convex for every $a \in X$.

Proposition 5.1.6. Let $X, \|\cdot\|$ be a normed space. For $\epsilon > 0$ define the open ball centred at 0 and with radius ϵ by $B_\epsilon = \{x \in X : \|x\| < \epsilon\}$. Then we have:

(a) B_ϵ is a convex set $\forall \epsilon > 0$.
(b) $\overline{B_\epsilon} = \{x \in X : \|x\| \leq \epsilon\}$, closed ball centred at 0 and with radius ϵ.
(c) $B_\epsilon = \epsilon B_1$.
(d) The family $\{B_\epsilon, \epsilon > 0\}$ is a base of convex open neighborhoods of vector 0.

Proof: (a) Let $x, y \in B_\epsilon$ and $0 \leq \lambda \leq 1$, we have
$$\|\lambda x + \left(1 - \lambda\right)y\| \leq \|\lambda x\| + \|\left(1 - \lambda\right)y\| = \lambda \|x\| + \left(1 - \lambda\right)\|y\|$$
$$< \lambda\epsilon + \left(1 - \lambda\right)\epsilon \subset$$
$$= \epsilon$$

so $\lambda x + \left(1 - \lambda\right)y \in B_\epsilon$.

(b) It is clear that $\{x \in X : \|x\| \leq \epsilon\}$ is a closed set and $\overline{B_\epsilon} \subset \{x \in X : \|x\| \leq \epsilon\}$. Since $\{x \in X : \|x\| \leq \epsilon\} = B_\epsilon \cup S_\epsilon$, with $S_\epsilon = \{x \in X : \|x\| = \epsilon\}$, it is enough to prove $S_\epsilon \subset \overline{B_\epsilon}$. Let $x \in S_\epsilon$ and let $\delta > 0$; consider the open ball $B\left(x, \delta\right)$ centred at x and with radius δ, then one can easily check that the vector $y = \alpha x$, with $\alpha \in \left]\left(1 - \delta\epsilon^{-1}\right)^+, 1\right[$, is in $B\left(x, \delta\right) \cap B_\epsilon$, where $\left(1 - \delta\epsilon^{-1}\right)^+ = Sup\left(1 - \delta\epsilon^{-1}, 0\right)$. Consequently every open neighborhood of x cuts B_ϵ, this proves that $x \in \overline{B_\epsilon}$.

(c) comes from the following equivalence:
$$x \in B_\epsilon \iff \epsilon^{-1}x \in B_1 \iff x \in \epsilon B_1$$

(d) Let V be any open neighborhood of 0 for the topology induced on X by the norm (Proposition **5.1.2.**); so there is $\epsilon > 0$ such that $B_\epsilon \subset V$, this shows that $\{B_\epsilon, \epsilon > 0\}$ is a base of convex open neighborhoods of vector 0 (see Definition **2.1.8**, Chapter **2**, for local base).∎

Corollary: In a normed space, every point has a base of convex open neighborhoods.

Proof: Let $B\left(x, \epsilon\right)$ be the open ball centered at x with radius ϵ. We have $B\left(x, \epsilon\right) = x + B_\epsilon$ and so $B\left(x, \epsilon\right)$ is convex (Example **5.1.5.**). Consequently $\{B\left(x, \epsilon\right), \epsilon > 0\}$ is base of convex open neighborhoods for x.∎

Proposition 5.1.7. Let $X, \|\cdot\|$ be a normed space. Let M be a vector subspace of X and suppose M closed. Consider the quotient X/M of X with respect to the equivalence relation $xRy \iff x - y \in M$. If u is in X/M

define: $\|u\| = \inf\{\|x\|, x \in u\}$

then $\|u\|$ is a norm on X/M, which will be denoted by $\|\cdot\|_Q$.

Proof: First let us point out that X/M is a vector space with operations:

$u, v \in X/M$, $u + v = x + y + M$, if $u = x + M, v = y + M$

$\lambda u = \lambda x + M$, for λ scalar.

It is easy to check that such operations are well defined, that is, depending on u, v and not on their representatives x, y. Note that the nul vector of X/M is the class M of the vector 0 of X. Now let us prove that $\|u\| = \inf\{\|x\|, x \in u\}$ is a norm:

(1) If $\|u\| = 0$, there is a sequence $x_n \in u$ such that $\|x_n\| \longrightarrow 0$. Let $a \in X$ with $u = a + M$, then there is $t_n \in M$ such that $x_n = a + t_n$; then $\|a + t_n\| \longrightarrow 0$; since M is closed we deduce that $-a \in M$ and then $u = M$, the nul vector of X/M.

(2) Let λ be a scalar and $u \in X/M$. Then $\|\lambda u\| = \inf\{\|t\|, t \in \lambda u\} = \inf\{\|\lambda x\|, \lambda x \in \lambda u\} = \inf\{|\lambda|\,\|x\|, \lambda x \in \lambda u\} = |\lambda|\inf\{\|x\|, \lambda x \in \lambda u\}$
$= |\lambda|\inf\{\|x\|, x \in u\} = |\lambda|\,\|u\|$.

(3) Let $u, v \in X/M$, then by the definition of $u + v$ we have:

$\|u + v\| = \inf\{\|x + y\|, x \in u, y \in v\} \leq \inf\{\|x\| + \|y\|, x \in u, y \in v\}$
$\leq \inf\{\|x\|, x \in u\} + \inf\{\|y\|, y \in v\}$. ∎

Theorem 5.1.8. Let $X, \|\cdot\|$ be a normed space and M a closed vector subspace of X. Consider the quotient X/M of Proposition **5.1.7.** and the canonical mapping $p : X \longrightarrow X/M$. Then we have:

The canonical mapping $p : X \longrightarrow X/M$ is open from $X, \|\cdot\|$ onto $X/M, \|\cdot\|_Q$.

Proof: Let $\epsilon > 0$ and consider the open balls:

$B_\epsilon = \{x \in X : \|x\| < \epsilon\} \subset X, \|\cdot\|$ and $G_\epsilon = \left\{u \in X/M : \|u\|_Q < \epsilon\right\}$

First we prove that $p(B_\epsilon) = G_\epsilon$; let $x \in B_\epsilon$ and put $u = p(x)$, then $\|u\|_Q \leq \|x\|$, from the defintion of $\|u\|_Q$ (Proposition **5.1.7.**), so $\|u\|_Q < \epsilon$, that is $u \in G_\epsilon$, and so $p(B_\epsilon) \subset G_\epsilon$. On the other hand let $u \in G_\epsilon$; since $\|u\|_Q < \epsilon$, there is $x \in u$ such that $\|x\| < \epsilon$ which gives $x \in B_\epsilon$ and $p(x) = u$, that is $G_\epsilon \subset p(B_\epsilon)$. Now consider an open set V of the nul vector 0 in $X, \|\cdot\|$, we have to prove that $p(V)$ is an open set of the nul vector M in $X/M, \|\cdot\|_Q$. Since V is open there is $\epsilon > 0$ such that $B_\epsilon \subset V$, then $p(B_\epsilon) = G_\epsilon \subset p(V)$, we deduce that $p(V)$ is an open set containing the nul vector M in $X/M, \|\cdot\|_Q$. In general, if $x \in X$ and if U be an open neighborhood of x in $X, \|\cdot\|$, there is an open set V of the nul vector 0 with $U = x + V$ (see lemma below); by the additivity of p we get $p(U) = p(x) + p(V)$; since $p(V)$ is an open set of the nul vector M in $X/M, \|\cdot\|_Q$ we deduce that $p(U)$ is an open neighborhood of $p(x)$. ∎

Lemma Let $E, \|\cdot\|$ be a normed space. Then the function $\varphi_t : E \longrightarrow E$, give by $\varphi_t(x) = x + t$, with t fixed in E, is an homeomorphism of E, with inverse $\varphi_t^{-1} = \varphi_{-t}$.

Proof: Straightforward. ∎

Proposition 5.1.9. Let X, Y be a normed spaces on the same field. For $(x, y) \in X \times Y$, the formula $|||(x,y)||| = ||x|| + ||y||$ defines a norm on $X \times Y$ whose induced topology coincides with the product topology on $X \times Y$.

Proof: The fact that $|||(x,y)|||$ is a norm is immediate. So we prove the second part of the proposition. Let $(s, t) \in X \times Y$, and let V_ϵ be the open ball centered at (s, t) with radius ϵ that is:

$V_\epsilon = \{(x,y) \in X \times Y : |||(x,y) - (s,t)||| < \epsilon\} = \{(x,y) \in X \times Y : ||x - s|| + ||y - t|| < \epsilon\}$.
It is clear that $\left\{x \in X : ||x - s|| < \dfrac{\epsilon}{2}\right\} \times \left\{y \in Y : ||y - t|| < \dfrac{\epsilon}{2}\right\} \subset V_\epsilon$, and the left
side is an open set for the product topology containing (s,t), so V_ϵ is an open for the
product topology. Conversely let W be an open neighborhood of (s,t) for the product topology of the form $W = \{x \in X : ||x - s|| < \alpha\} \times \{y \in Y : ||y - t|| < \beta\}$;putting
$\gamma = Min\,(\alpha,\beta)$ we get $V_\gamma = \{(x,y) \in X \times Y : ||x - s|| + ||y - t|| < \gamma\} \subset W$, and
since V_γ is open for the norm topology and contains (s,t), we deduce that W is
neighborhood of (s,t) for the norm topology.∎

Proposition 5.1.10. Let $X, ||\cdot||$ be a normed space on the field $\mathbb{K} = \mathbb{R}$ or $\mathbb{C}$. Then
the norm topology on X is compatible with the algebraic vector structure of X,
which means that the functions:

$(x,y) \longrightarrow x + y$ from $X \times X$ into X

$(\lambda, x) \longrightarrow \lambda x$ from $\mathbb{K} \times X$ into X

are continuous with the product topologies on $X \times X$ and $\mathbb{K} \times X$.

Proof: Let $x,y \in X$ and $\lambda \in \mathbb{K}$. Let us consider sequences $(x_n), (y_n) \subset X$,
$(\lambda_n) \subset \mathbb{K}$ such that $x_n \longrightarrow x, y_n \longrightarrow y$ and $\lambda_n \longrightarrow \lambda$ as $n \longrightarrow \infty$. Then we have:

$||(x_n + y_n) - (x + y)|| \leq ||x_n - x|| + ||y_n - y|| \longrightarrow 0, n \longrightarrow \infty$

$||\lambda_n x_n - \lambda x|| = ||\lambda_n (x_n - x) + (\lambda_n - \lambda) x||$

$\leq |\lambda_n| \, ||x_n - x|| + |\lambda_n - \lambda| \, ||x|| \longrightarrow 0, n \longrightarrow \infty.$∎

Corollary 1: Let $X, ||\cdot||$ be a normed space. Fix an $x_0 \in X$ and a scalar $\lambda_0 \neq 0$,
then the functions $x \longrightarrow x + x_0$ and $x \longrightarrow \lambda_0 x$ are homeomorphisms of X.

Corollary 2: In a normed space $X, ||\cdot||$, the neighborhoods of a vector x are of the
form $x + V$, where V is a neighborhoods of 0.

Proofs are left as exercises to the reader.∎

Definition 5.1.11. Let (x_n) be an infinite sequence in a normed space $X, ||\cdot||$, and
put $s_n = x_1 + x_2 + ... + x_n$.

    (a) We say that the series s_n converges to $s \in X$ if $\underset{n}{Lim} \, ||s_n - s|| = 0$.

    We denote $s = \sum_n x_n$.

    (b) We say that the series s_n converges absolutely if $\sum_n ||x_n|| < \infty$.

The relation between these two convergence types is the best one when the normed
space X is complete according to the following definition:

Definition 5.1.12. A Banach space is a normed space which is complete with
respect to the metric induced by the norm. (see Proposition **5.1.2.** for the metric
induced by a norm).

Proposition 5.1.13. A normed space $X, ||\cdot||$ is a Banach space if and only if every
absolutely convergent series is convergent.

Proof: Suppose that X is a Banach space and let (x_n) be an infinite sequence in
X with $\sum_n ||x_n|| < \infty$, and $s_n = x_1 + x_2 + ... + x_n$; we prove that (s_n) is Cauchy.

Indeed we have for $m \geq n$ $||s_m - s_n|| = \left\|\sum_{n+1}^m x_k\right\| \leq \sum_{n+1}^m ||x_k|| \leq \sum_{n+1}^\infty ||x_k|| \longrightarrow 0$,
since this last sum is the remainder of a convergent positive series.

So $\|s_m - s_n\| \longrightarrow 0, m, n \longrightarrow \infty$, and (s_n) is Cauchy. As X is complete, s_n converges to some $s \in X$.

Conversely assume the condition of the proposition for the space $X, \|\cdot\|$. We have to prove that X is complete. Let (x_n) be Cauchy in X, so for each integer $k \geq 1$ there is $N_k \geq 1$ such that $\forall m, n \geq N_k, \|x_n - x_m\| < \dfrac{1}{2^k}$.

Let n_k be the subsequence of integers given by:

$\quad n_1 = N_1$ and for $k \geq 2, n_k = Max\,(N_k, n_{k-1} + 1)$

Then we have $1 \leq n_1 < n_2 < ... < n_k \longrightarrow \infty$ and $\left\|x_{n_{k+1}} - x_{n_k}\right\| < \dfrac{1}{2^k}, \forall k \geq 1$,

from this we deduce that the series $x_{n_1} + \sum\limits_{n+1}^{\infty} \left(x_{n_{k+1}} - x_{n_k}\right)$ is absolutely convergent and so convergent by the condition of the proposition.

But $x_{n_1} + \sum\limits_{k=1}^{m-1} \left(x_{n_{k+1}} - x_{n_k}\right) = x_{n_m}$, then the subsequence (x_{n_m}) converges to some $x \in X$ and so the sequence (x_n) itself converges to x, since $\|x_n - x\| \leq \|x_n - x_{n_m}\| + \|x - x_{n_m}\| \longrightarrow 0, n, m \longrightarrow \infty.\blacksquare$

Examples 5.1.14. (1) The space $\mathbb{R}^n$ is a Banach space for the euclidean norm. We will see later that $\mathbb{R}^n$ is a Banach space for every norm.

(2) The l_p spaces, $1 \leq p \leq \infty$ are Banach spaces $\forall 1 \leq p \leq \infty$.(see Example **5.1.3.** $(2), (3)$ for the definition of these spaces). We prove the property for $1 \leq p < \infty$, and leave the easier case $p = \infty$ as exercise.

Let (x_n) be Cauchy in l_p, with $x_n = (\alpha_{n,k})_{k \geq 1}$. From the definition of the norm in l_p, we have $\|x_n - x_m\|_p = \left(\sum\limits_k |\alpha_{n,k} - \alpha_{m,k}|^p\right)^{\frac{1}{p}}$, so for $\epsilon > 0$ there is $N \geq 1$ such that $n, m \geq N \implies \sum\limits_k |\alpha_{n,k} - \alpha_{m,k}|^p < \epsilon^p$.

We deduce that $\forall L \geq 1, \sum\limits_{k=1}^{L} |\alpha_{n,k} - \alpha_{m,k}|^p < \epsilon^p$, in particular $|\alpha_{n,r} - \alpha_{m,r}|^p < \epsilon^p$, for each $r \geq 1$; this means that the sequence $(\alpha_{n,r})_{n \geq 1}$ is Cauchy in $\mathbb{R}$.

Put $\alpha_r = \lim\limits_n \alpha_{n,r}$, we prove that the vector $x = (\alpha_r)_{r \geq 1}$ is in l_p, and that x_n converges to x in l_p. For each $L \geq 1$, and every $n \geq N$ $\lim\limits_{m \longrightarrow \infty} \sum\limits_{k=1}^{L} |\alpha_{n,k} - \alpha_{m,k}|^p = \sum\limits_{k=1}^{L} |\alpha_{n,k} - \alpha_k|^p \leq \epsilon^p$. On the other hand we have:

$\forall n \geq N, \sum\limits_k |\alpha_{n,k} - \alpha_k|^p = \lim\limits_{L \longrightarrow \infty} \sum\limits_{k=1}^{L} |\alpha_{n,k} - \alpha_k|^p \leq \epsilon^p$, and this proves that $x_n - x \in l_p$ with $\|x_n - x\|_p \leq \epsilon, \forall n \geq N$, that is x_n converges to x.

(3) Let X be a topological space, the space $C_b(x)$ of all continuous and bounded functions $f : X \longrightarrow \mathbb{R}$, with the norm $\|f\|_\infty = \underset{x \in X}{Sup}\,|f(x)|$ is a Banach space (see Example **5.1.3.** (4)). This comes from the fact that a uniform limit of a sequence of continuous bounded functions, is a continuous bounded function.

(4) Let X be a normed space and M a closed subspace of X. Let the quotient X/M be equipped with the quotient norm $\|u\| = \inf\{\|x\|, x \in u\}$ (Proposition **5.1.7.**). If X is a Banach space then X/M is a Banach space. To see this we

use the criteria of Proposition **5.1.13**. Let (u_n) be an absolutely convergent series in X/M, we prove that (u_n) is convergent. From the definition of $\|u_n\|$, we get: $\forall n \geq 1, \exists x_n \in u_n : \|x_n\| \leq 2\|u_n\|$. So the series (x_n) is absolutely convergent in X, then it is convergent since X is Banach, let x be its limit. Put $u = x + M$, the class of x, then we have $\left\|\sum_1^N u_n - u\right\| \leq \left\|\sum_1^N x_n - x\right\| \longrightarrow 0, N \longrightarrow \infty$. From this it comes that $u = \sum_n u_n$, which proves the convergence of the series (u_n).

2. Exercises

82. Let X be a vector space. A metric d on X is said to be translation invariant if $d(x + a, y + a) = d(x, y), \forall a, x, y \in X$.

(a) If X is a normed space then the metric induced by the norm is translation invariant.

(b) Let d be a translation invariant metric on X and put $\|x\| = d(0, x)$.
Prove that $\|-x\| = \|x + y\|, \leq \|x\| + \|y\|$, and $\|x\| = 0 \iff x = 0$.

83. Let $C^1[0, 1]$ be the space of continuous functions on $[0, 1]$ having continuous first derivative (functions of class C^1). For $f \in C^1[0, 1]$, define:
$$\|f\|_1 = \sup_{t \in [0,1]} (|f(t)| + |f'(t)|)$$
$$\|f\|_2 = \sup_{t \in [0,1]} |f(t)| + \sup_{t \in [0,1]} |f'(t)|$$
Prove that $\|f\|_1$ and $\|f\|_2$ are norms on $C^1[0, 1]$ with $\|f\|_1 \leq \|f\|_2$.
Give an example of a function $f \in C^1[0, 1]$ with $\|f\|_1 < \|f\|_2$.

84. A semi-norm on a vector space X is a function $p : X \longrightarrow [0, \infty)$ satisfying:
(1) $p(x + y) \leq p(x) + p(y), \forall x, y \in X$.
(2) $p(\lambda x) = |\lambda| p(x)$, for every scalar λ.
(a) Prove that $N = \{x : p(x) = 0\}$ is a vector subspace of X.
(b) Define on the quotient space X/N the function $\|\cdot\|$ by:
$u \in X/N, \|u\| = p(x)$, if $u = x + N$.
Prove that $\|\cdot\|$ is a well defined norm on X/N.

85. Let X be a normed space and $x, y \in X$. If $z = x + y$ then for every open neighborhood W of z, there is an open neighborhood U of x and an open neighborhood V of y such that $U + V \subset W$.

86. Let X be a normed space and let M be a vector subspace of X.
Prove that the closure $\overline{M}$ of M is also a subspace of X.

87. Let X be a normed space and let E, F be subsets of X such that E is compact and F closed. Prove that $E + F$ is closed.

88. Let X, Y be normed spaces and let the product $X \times Y$ be equipped with the product norm (Proposition **5.1.9.**). Prove that $X \times Y$ is a Banach space if and only if X and Y are Banach spaces.

89. Let X be a normed space and let A be a subset of X. Prove that for every scalar α and every vector x, we have:

$$\overset{o}{\alpha A} = \alpha \overset{o}{A}$$
$$\overline{\alpha A} = \alpha \overline{A}$$
$$\overset{o}{x + A} = x + \overset{o}{A}$$
$$\overline{x + A} = x + \overline{A}$$

90. Let us consider the following spaces of sequences:

$$c_0 = \left\{ a = (a_n) \in \mathbb{R}^\mathbb{N} : \lim_n a_n = 0 \right\}$$
$$d_0 = \left\{ a = (a_n) \in \mathbb{R}^\mathbb{N} : \exists N \geq 1, a_n = 0, \forall n \geq N \right\}$$

(a) Prove that c_0, d_0 are vector subspaces of $\mathbb{R}^\mathbb{N}$ and we have:

$$d_0 \subset l_p \subset c_0 \subset l_\infty$$

(b) Prove that d_0 is dense in l_p for $\|\cdot\|_p$ and d_0 is dense in c_0 for $\|\cdot\|_\infty$.

3. Linear Bounded Operators

A linear operator from a normed space X into a normed space Y is a linear mapping from X into Y.

Definition 5.3.1. A linear operator T from a normed space X into a normed space Y is said to be bounded if there is a constant $M > 0$ such that:

$$\|T(x)\| \leq M.\|x\|, \forall x \in X$$

This definition means that if B is a bounded subset of X, the set $\{T(x), x \in B\}$ is bounded in Y. For instance if $B = \{x : \|x\| \leq 1\}$ then $\|T(x)\| \leq M, \forall x \in B$.

Examples 5.3.2. Let the space $C[0,1]$ of continuous functions $f : [0,1] \longrightarrow \mathbb{R}$, be equipped with the uniform norm $\|\cdot\|_\infty$. Define $T : C[0,1] \longrightarrow \mathbb{R}$ by $T(f) = \int_0^1 f(x)\, dx$ (Riemann integral), then it is clear that $|T(f)| \leq \|f\|_\infty$ and T is bounded with the choice $M = 1$.

Proposition 5.3.3. Let T be a bounded operator from X into Y. Define:

$$\|T\| = \sup \left\{ \frac{\|T(x)\|}{\|x\|} : x \in X, x \neq 0 \right\}$$
$$m_1 = \sup \{ \|T(x)\| : x \in X, \|x\| = 1 \}$$
$$m_2 = \sup \{ \|T(x)\| : x \in X, \|x\| < 1 \}$$
$$m_3 = \sup \{ \|T(x)\| : x \in X, \|x\| \leq 1 \}$$

Then $m_1 = m_2 = m_3 = \|T\| < \infty$ and we have:

$$\|T(x)\| \leq \|T\| \|x\|, \forall x \in X$$

Proof: First T bounded operator $\implies m_i < \infty, i = 1, 2, 3$, and $\|T\| < \infty$; next we have from the definition, $m_1 \leq \|T\|$. On the other hand if $x \neq 0$, then $\left\| \dfrac{x}{\|x\|} \right\| = 1$, and $\left\| T \left(\dfrac{x}{\|x\|} \right) \right\| = \dfrac{\|T(x)\|}{\|x\|} \leq m_1$, this yields $\|T\| \leq m_1$, so $\|T\| = m_1$.

Since $\{x : \|x\| = 1\} \subset \{x : \|x\| \leq 1\}$, we get $m_1 \leq m_3$; on the other hand for $\|x\| \leq 1, x \neq 0$, we have $\left\| T \left(\dfrac{x}{\|x\|} \right) \right\| = \dfrac{\|T(x)\|}{\|x\|} \leq m_1$, whence $\|T(x)\| \leq \|x\| m_1 \leq m_1$.

Taking supremum over $\{x : \|x\| \leq 1\}$ we get $m_3 \leq m_1$, so $m_1 = m_3$. By the same trick we obtain $m_1 = m_2$. Finally it is clear that $\|T(x)\| \leq \|T\| \|x\|$, $\forall x \in X.\blacksquare$

Theorem 5.3.4. The following properties are equivalent for a linear operator T from X into Y:

    (a) T is bounded.

    (b) T is uniformly continuous.

    (c) T is continuous at a point $x_0 \in X$.

Proof: $(a) \implies (b)$

We have $\|T(x) - T(x')\| = \|T(x - x')\| \leq \|T\| \|x - x'\|$, $\forall x, x' \in X$, so if $\epsilon > 0$ then we have $\|x - x'\| < \epsilon \|T\|^{-1} \implies \|T(x) - T(x')\| < \epsilon$.

$(b) \implies (c)$ is trivial.

$(c) \implies (a)$

By (c) there is $\sigma > 0$ such that $\|x - x_0\| < \sigma \implies \|T(x) - T(x_0)\| < 1$. Now if $\|x\| < 1$, we get $\|\sigma x + x_0 - x_0\| < \sigma$, and since $x = \dfrac{1}{\sigma}(\sigma x + x_0 - x_0)$, we deduce $\|T(x)\| = \dfrac{1}{\sigma}\|T(\sigma x + x_0) - T(x_0)\| < \dfrac{1}{\sigma}$. From Proposition **5.3.3**, $\|T\| = \sup\{\|T(x)\| : x \in X, \|x\| < 1\}$, and then $\|T\| \leq \dfrac{1}{\sigma}$, this yields T bounded.$\blacksquare$

We denote by $B(X, Y)$ the set of linear bounded operators from a normed space X into a normed space Y, on the same field $\mathbb{K}$ of scalars. Let $S, T \in B(X, Y)$ and $\alpha \in \mathbb{K}$, we define for $x \in X$:

    $(S + T)(x) = S(x) + T(x)$

    $(\alpha T)(x) = \alpha T(x)$

Then we have:

Proposition 5.3.5. $B(X, Y)$ is a vector space with these defined operations. Moreover the function $T \longrightarrow \|T\|$ is a norm on $B(X, Y)$.

Proof: It is immediate that $B(X, Y)$ is a vector space, the null vector being the operator T with $T(x) = 0, \forall x \in X$. To see that $T \longrightarrow \|T\|$ is a norm, let $S, T \in B(X, Y)$, then we have $\|S + T\| = \sup\{\|S(x) + T(x)\|, \|x\| = 1\} \leq \sup\{\|S(x)\| + \|T(x)\|, \|x\| = 1\} \leq \sup\{\|S(x)\|, \|x\| = 1\} + \sup\{\|T(x)\|, \|x\| = 1\} = \|S\| + \|T\|$. Likewise $\|\alpha T\| = |\alpha| \|T\|, \alpha \in \mathbb{K}$. Finally, since $\|T(x)\| \leq \|T\| \|x\|$, we have $\|T\| = 0 \implies T(x) = 0, \forall x \in X.\blacksquare$

Theorem 5.3.6. If Y is a Banach space, $B(X, Y)$ is a Banach space.

Proof: Let (T_n) be Cauchy in $B(X, Y)$. For each $x \in X$ we have $\|T_n(x) - T_m(x)\| \leq \|T_n - T_m\| \|x\|, \forall n, m \geq 1$, so $(T_n(x))$ is Cauchy in Y and since Y is Banach $\lim_n T_n(x)$ exists in Y; we denote this limit by $T(x)$.

The mapping so defined from X into Y is linear.

Indeed, for each $n \geq 1$ we have $\|T(x + y) - (T(x) + T(y))\| \leq$
$\|T(x + y) - T_n(x + y)\| + \|T_n(x + y) - (T(x) + T(y))\| \leq$
$\|T(x + y) - T_n(x + y)\| + \|T_n(x) - T(x)\| + \|T_n(y) - T(y)\| \longrightarrow 0, n \longrightarrow \infty$. So we get $T(x + y) = T(x) + T(y)$. Similarly $T(\alpha x) = \alpha T(x), \alpha \in \mathbb{K}, x \in X$.

It remains to prove that $T \in B(X, Y)$ and that $\|T_n - T\| \longrightarrow 0$. If $\epsilon > 0$ there is $N_\epsilon \geq 1$: $n, m \geq N_\epsilon \implies \|T_n - T_m\| < \epsilon$. For $n \geq N_\epsilon$, we have $\|T_n(x) - T(x)\| = \lim_m \|T_n(x) - T_m(x)\| \leq \limsup_m \|T_n - T_m\| \|x\| \leq \epsilon \|x\|$.

Consequently if $n \geq N_\epsilon$, we get $\|T_n(x) - T(x)\| \leq \epsilon \|x\|, \forall x \in X$, and this yields $T_n - T \in B(X,Y)$ and $T \in B(X,Y)$; on the other hand for $n \geq N_\epsilon$
$\|T_n - T\| = \sup\{\|T_n(x) - T(x)\|, \|x\| = 1\} \leq \epsilon$, that is $\|T_n - T\| \longrightarrow 0.\blacksquare$

Definition 5.3.7. Let X be a normed space on the field $\mathbb{K} = \mathbb{R}$ or $\mathbb{C}$. A linear functional or linear form on X is a linear operator from X into $\mathbb{K}$. We denote by X^*, instead of $B(X,\mathbb{K})$, the Banach space of continuous linear functionals on X. The space X^* is called the dual space of X. Likewise we denote by X^{**} the dual space of X^* and call it the bidual space of X.

One of the famous theorems related to the space linear functionals on X , useful in many applications, is the following:

Theorem 5.3.8.   (Hahn-Banach)
Let X be a vector space on the field $\mathbb{K} = \mathbb{R}$ or $\mathbb{C}$, and let $p : X \longrightarrow [0,\infty)$ be a seminorm on X, that is p satisfies:
$(1)\, p(x+y) \leq p(x) + p(y), \forall x, y \in X.$
$(2)\, p(\alpha x) = |\alpha| p(x), \forall \alpha \in \mathbb{K}, \forall x \in X.$
Let M be a subspace of X and $g : M \longrightarrow \mathbb{K}$, a linear form on M such that:

$$|g(y)| \leq p(y), \forall y \in M$$

Then g can be extended to a linear functional $f : X \longrightarrow \mathbb{K}$, on X such that:

$$|f(x)| \leq p(x), \forall x \in X$$

Proof: see [] $\blacksquare$

For applications, the following corollaries are the most useful:

Corollary 1: Let X be a vector space on the field $\mathbb{K} = \mathbb{R}$ or $\mathbb{C}$
and let $p : X \longrightarrow [0,\infty)$ be a seminorm on X, then for each $a \in X$, there is a linear functional f on X such that $f(a) = p(a)$ and $|f(x)| \leq p(x), \forall x \in X$.

Proof: If $a = 0$ take $f \equiv 0$. If $a \neq 0$, consider the subspace M generated by a, that is $M = \mathbb{K}.a$, and define $g : M \longrightarrow \mathbb{K}$ by $g(\lambda a) = \lambda p(a)$, $\lambda \in \mathbb{K}$. Then g is a linear functional on M which satisfies the conditions of Theorem **5.3.8.**, since we have $|g(\lambda a)| = |\lambda| p(a) = p(\lambda a)$. So there is a linear functional $f : X \longrightarrow \mathbb{K}$, on X such that $f(y) = g(y), \forall y \in M$, and $|f(x)| \leq p(x), \forall x \in X$. But $a \in M$ and $f(a) = g(a) = p(a).\blacksquare$

The next corollary shows that in a normed space, continuous linear functionals exist in profusion:

Corollary 2: Let X be a normed space, then for each $a \in X$, there is $f \in X^*$ such that $f(a) = \|a\|$, $|f(x)| \leq \|x\|, \forall x \in X$ and $\|f\| = 1$.

Proof: Apply corollary **1** with $p(x) = \|x\|$: for each $a \in X$, there is a linear functional f on X such that $f(a) = \|a\|$ and $|f(x)| \leq \|x\|, \forall x \in X$. It is clear that such functional is continuous and $\|f\| \leq 1$. On the other hand we have
$\|f\| = \sup\left\{\dfrac{|f(x)|}{\|x\|} : x \in X, x \neq 0\right\} \geq \dfrac{|f(a)|}{\|a\|} = 1$, so we get $\|f\| = 1.\blacksquare$

Corollary 3: Let X be a normed space and let M be a subspace of X.
If $g : M \longrightarrow \mathbb{K}$, is a continuous linear functional on M, i.e $g \in M^*$, there is an extension f of g to X with $f \in X^*$ and $\|f\| = \|g\|$.

Proof: Put $p(x) = \|g\| \|x\|$. Then since g is continuous we have $|g(x)| \leq \|g\| \|x\| = p(x)$ on M. Apply Theorem **5.3.8** to get an extension f of g to X with $|f(x)| \leq \|g\| \|x\| = p(x)$. This shows that $f \in X^*$ and $\|f\| \leq \|g\|$. But we have also $\|f\| = \sup\{|f(x)| : x \in X, \|x\| \leq 1\} \geq \sup\{|f(y)| : y \in M, \|y\| \leq 1\} = \sup\{|g(y)| : y \in M, \|y\| \leq 1\} = \|g\|$. $\blacksquare$

Definition 5.3.9. Let X, Y, E be normed spaces on the same field $\mathbb{K} = \mathbb{R}$ or $\mathbb{C}$. A mapping $T : X \times Y \longrightarrow E$ is bilinear if:

$\forall x, x' \in X, \forall y, y' \in Y, \forall \alpha, \beta \in \mathbb{K}$

$T(\alpha x + \beta x', y) = \alpha T(x, y) + \beta T(x', y)$

$T(x, \alpha y + \beta y') = \alpha T(x, y) + \beta T(x, y')$

In other words T is bilinear if the following mappings are linear:

$x \longrightarrow T(x, y), y$ fixed in Y

$y \longrightarrow T(x, y), x$ fixed in X

Note also that if T is bilnear then $T(\alpha x, \beta y) = \alpha \beta T(x, y)$.

Proposition 5.3.10. Let $T : X \times Y \longrightarrow E$ be a bilinear mapping, $X \times Y$ equipped with the product topology. The following properties are equivalent:

(a) T is continuous on $X \times Y$

(b) T is continuous at $(0, 0)$

(c) There is a constant $M > 0$ such that:

$\|T(x, y)\| \leq M \|x\| \|y\|, \forall (x, y) \in X \times Y$

Proof: First let us recall that the product topology on $X \times Y$ is induced by the norm $\||(x, y)\|| = \|x\| + \|y\|$ (Proposition **5.1.9**).

$(a) \implies (b)$ is immediate

$(b) \implies (c)$: Let B be the unit open ball in E. Since $T(0, 0) = 0$, condition (b) shows that $T^{-1}(B)$ is an open neighborhood of $(0, 0)$ in $X \times Y$, so there is $c > 0$ such that $\|x\| < c, \|y\| < c$ implies $\|T(x, y)\| \leq 1$. Let $0 < d < c$ and $x \neq 0, y \neq 0$, then we have $\left\| d\dfrac{x}{\|x\|} \right\| < c$ and $\left\| d\dfrac{y}{\|y\|} \right\| < c$; consequently $\left\| T\left(d\dfrac{x}{\|x\|}, d\dfrac{y}{\|y\|}\right) \right\| = \dfrac{d^2}{\|x\| \|y\|} T(x, y) \leq 1$ and condition (c) is satisfied with the constant $M = \dfrac{1}{d^2}$.

$(c) \implies (a)$: Let $(a, b) \in X \times Y$, we have by the bilinearity $T(x, y) - T(a, b) = T(x - a, y) + T(a, y - b)$; taking the norms we get from (c) $\|T(x, y) - T(a, b)\| \leq \|T(x - a, y)\| + \|T(a, y - b)\| \leq M \|x - a\| \|y\| + M \|a\| \|y - b\|$.

But $\|y\| \leq \|y - b\| + \|b\|$, consequently $\|T(x, y) - T(a, b)\| \leq M(\|x - a\| \|y - b\| + \|x - a\| \|b\| + \|a\| \|y - b\|) \longrightarrow 0$ if $(x, y) \longrightarrow (a, b)$. $\blacksquare$

4. Exercises

91. Let X, Y be normed spaces and $T : X \longrightarrow Y$ a linear operator.
Prove that T is bounded iff $T^{-1}\{y : \|y\| \leq 1\}$ has nonempty interior.

92. Let $\lambda = (\lambda_n) \in l_1$ and define the linear operator φ_λ on the space c_0 by:

$a = (a_n) \in c_0, \varphi_\lambda(a) = \sum_n \lambda_n . a_n$

Prove that φ_λ is bounded and $\|\varphi_\lambda\| = \sum_n |\lambda_n| = \|\lambda\|_{l_1}$.

93. Let $\lambda = (\lambda_n) \in l_\infty$ and define the linear operator ψ_λ on the space l_1 by:
$$b = (b_n) \in l_1, \ \psi_\lambda(b) = \sum_n \lambda_n . b_n$$
Prove that ψ_λ is bounded and $\|\psi_\lambda\| = \sup_n |\lambda_n| = \|\lambda\|_\infty$.

94. Let X,Y be normed spaces and $T : X \longrightarrow Y$ a bounded linear operator.
(a) Let $M = \{x \in X : T(x) = 0\}$ be the kernel of T. Prove that M is a closed subspace of X .
(b) Define $\widetilde{T} : X/M \longrightarrow Y$ by $\widetilde{T}(x + M) = T(x)$. Prove that $\widetilde{T}$ is well defined, linear and injectif.
(b) We equip X/M with the quotient norm (Proposition **5.1.7**)
   Prove that $\left\|\widetilde{T}\right\| = \|T\|$.

95. Let X be a normed space and let M be a closed subspace of X. We equip X/M with the quotient norm, and consider the canonical projection $\pi : X \longrightarrow X/M$. Prove that π is linear continuous and $\|\pi\| \leq 1$.

96. Let X be a compact topological space and let $C(X)$ be the space of all continuous functions $f : X \longrightarrow \mathbb{R}$, equipped with the uniform norm.
Let $T : C(X) \longrightarrow \mathbb{R}$, be a linear functional satisfying the positivity property:
$$f \in C(X), f \geq 0 \implies T(f) \geq 0$$
(a) Prove that: $f \leq g \implies T(f) \leq T(g)$ and $|T(f)| \leq T(|f|)$.
(b) Prove that T is continuous and $\|T\| = T(1)$,
   where 1 is the constant function equal to 1.

97. Let X be a normed space and let S be a dense subspace of X. Let $f : S \longrightarrow \mathbb{R}$ be a linear continuous functional on S. Prove, without using Hahn-Banach theorem, that there exists $g \in X^*$ such that $g(x) = f(x), \forall x \in S$ and $\|g\| = \|f\|$.

98. Let X be a normed space and let M be a closed subspace of X. Fix $z \notin M$ and consider the subspace S generated by M and z, that is:
$$S = \{x + \alpha z, x \in M, \alpha \in \mathbb{K}\}$$
Define $f : S \longrightarrow \mathbb{K}$ by $f(x + \alpha z) = \alpha$.
(a) Prove that $f \in S^*$ and $\|f\| = \dfrac{1}{d}$, where $d = d(z, M)$, is the distance from z to M (Definition **3.1.14**(b) Chapter 3); note that $d > 0$ since M is closed (see Exercise **41** Chapter 3).
(b) Prove that S is closed.
(c) Prove that f has an extension g to all of X
   with $\|g\| = \|f\|$ and $g(x) = 0, \forall x \in M$.
(Apply corollary 3 of Hahn-Banach theorem).

99. Let (T_n) be a sequence in $B(X,Y)$ and suppose there is $T \in B(X,Y)$ such that $\|T_n - T\| \longrightarrow 0, n \longrightarrow \infty$. Let (x_n) be a sequence in X with $x_n \to x$, prove that $T_n(x_n) \longrightarrow T(x)$ in Y.

100. (a) Prove that the l_p spaces $1 \leq p < \infty$ are separable.
   (b) Prove that the space c_0 is separable for the norm $\|\cdot\|_\infty$.
   (c) Prove that $c_0^* = l_1$ and $l_1^* = l_\infty$
   (this means: there is a linear bijection $\varphi : l_1 \to c_0^*$ (resp. $\psi : l_\infty \to l_1^*$) such that $\|\varphi(x)\| = \|x\|$ (resp. $\|\psi(x)\| = \|x\|$).).

As for the separability of Banach spaces we have the following theorem whose proof can be found in [Linear Op.]

Theorem 5.4.1. Let X be a Banach space such that the dual X^* is separable, then X is separable.

5. Normed Spaces of Finite Dimension

Definition 5.5.1. Let X,Y, be normed spaces. A mapping $\varphi : X \longrightarrow Y$ is an isomorphism from X onto Y if:

    (i) φ is linear and bijective

    (ii) φ and its inverse φ^{-1} are continuous.

condition (i) implies that φ^{-1} is linear.

Using Proposition **5.3.3**, we rephrase this definition as follows:

Proposition 5.5.2. A linear bijective mapping $\varphi : X \longrightarrow Y$ is an isomorphism of normed spaces if and only if there are positive constants α, β such that:
$$\|\varphi (x)\| \leq \alpha \|x\| , \forall x \in X$$
$$\|\varphi^{-1} (y)\| \leq \beta \|y\| , \forall y \in Y$$

Proof: Apply Theorem **5.3.4.**■

Definition 5.5.3. Let X be a normed space equipped with two norms $\|\cdot\|_1 , \|\cdot\|_2$. We say that $\|\cdot\|_1 , \|\cdot\|_2$ are equivalent if there is $\alpha > 0, \beta > 0$ such that:
$$\|x\|_2 \leq \alpha \|x\|_1 \text{ and } \|x\|_1 \leq \beta \|x\|_2 \quad \forall x \in X$$

Two remarks are in order from this definition:

(1) The equivalence of the norms $\|\cdot\|_1 , \|\cdot\|_2$ means that the identity mapping from $X, \|\cdot\|_1$ into $X, \|\cdot\|_2$ is an isomorphism.

(2) Two equivalent norms on X induce the same topology on X.

Norms on spaces of finite dimension have special properties.

Proposition 5.5.4. Suppose $\mathbb{R}^n$ endowed with the product topology and let p be any norm on $\mathbb{R}^n$, then p is continuous.

Proof: Note that the product topology is induced by the euclidean norm of $\mathbb{R}^n$

$\|x\|_2 = \left(\sum_{i=1}^{n} |x_i|^2 \right)^{\frac{1}{2}}$. Let $e_1, e_2, ..., e_n$ be the canonical base of $\mathbb{R}^n$ and let $x = (x_1, x_2, ..., x_n), y = (y_1, y_2, ..., y_n)$ be vectors in $\mathbb{R}^n$.

Then we have $x - y = \sum_{i=1}^{n} (x_i - y_i).e_i$ and $|p(x) - p(y)| \leq p(x - y)$

$\leq \sum_{i=1}^{n} |x_i - y_i| .p(e_i) \leq n. \left(\sum_{i=1}^{n} p(e_i) \right) \|x - y\|_2$, whence the continuity of p.

(see the inequalities in Example **5.1.3.** (1)).■

Theorem 5.5.5. All the norms on $\mathbb{R}^n$ are equivalent.

Proof: Let p be an arbitrary norm on $\mathbb{R}^n$, it is enough to prove that p is equivalent to the euclidean norm. By Proposition **5.5.4** p is continuous, so it is bounded on the unit sphere $S = \{x : \|x\|_2 = 1\}$. Let $\alpha = \inf_{x \in S} p(x)$, $\beta = \sup_{x \in S} p(x)$, then we have $\alpha > 0$ (why?), and so we deduce that $\alpha \|x\|_2 \leq p(x) \leq \beta \|x\|_2 , \forall x \in X.$■

Corollary: All the norms on a finite dimensional vector space are equivalent.

Theorem 5.5.6. Let X be a normed space and let $\varphi : \mathbb{R}^n \longrightarrow X$ be a linear bijection. Then φ is an isomorphism of normed spaces.

Proof: Since φ is linear and bijective, it follows that the dimension of X is equal to n. Let p be an arbitrary norm on X, by the linearity of φ $p \circ \varphi$ is a norm on $\mathbb{R}^n$ which is equivalent to the euclidean norm (Theorem **5.5.5**); consequently there is a constant α such that $p \circ \varphi(x) \leq \alpha \|x\|_2 , \forall x \in \mathbb{R}^n$, so φ is continuous. A similar argument gives the continuity of φ^{-1}.∎

Theorem 5.5.7. Every finite dimensional normed space is a Banach space.

Proof: Let n be the dimension of X and let $\varphi : \mathbb{R}^n \longrightarrow X$ be a linear bijection. Then φ is an isomorphism of normed spaces, by the preceding theorem, so φ is bicontinuous. Since $\mathbb{R}^n$ is complete we deduce that X is complete.∎

6. Exercises

101. Prove that every linear mapping from $\mathbb{R}^n$ into a normed space is continuous.

102. Let X be a vector space with two norms p, q such that $p \leq \alpha q$ for some $\alpha > 0$
    (a) Prove that X, q separable $\implies X, p$ separable.
    (b) Prove that l_1 is separable for the norm $\|\cdot\|_\infty$.

102. Let $T : X \longrightarrow X$ be a linear bounded operator on a normed space $X, \|\cdot\|$. Define on X the function: $\||x\|| = \|x\| + \|T(x)\|$.
    (a) Prove that $\||\cdot\||$ is a norm equivalent to the norm $\|\cdot\|$.
    (b) Prove that $X, \||\cdot\||$ is a Banach space iff $X, \|\cdot\|$ is a Banach space.

7. Linear Bounded Operators in Banach Spaces

Linear bounded operators in Banach spaces have remarkable properties, using in an essential way the fact that such spaces are Baire spaces (see Chapter 3). Let us start with the following fundamental lemma.

Lemma 5.7.1. Let X, Y be Banach spaces and let $T : X \longrightarrow Y$ be a bounded operator. Suppose that T is onto and consider the open unit ball $A = \{x \in X : \|x\| < 1\}$ in X. Then there is $r > 0$ such that $B = \{y \in Y : \|y\| < r\} \subset T(A)$. In other words $T(A)$ is an open neighborhood of 0 in Y.

Proof: Let $S_n = \left\{x \in X : \|x\| < \dfrac{1}{2^n}\right\}, n \geq 1$.

We have $X = \bigcup_k k.S_1$, because if $x \in X$ there is $k \geq 1$ such that $\|x\| < \dfrac{k}{2}$, i.e $x \in k.S_1$; we deduce that $T(X) = Y = \bigcup_k kT(S_1) = \bigcup_k \overline{kT(S_1)}$.

Note that $\overline{k.T(S_1)} = k.\overline{T(S_1)}$ $\overline{k.T(S_1)}^o = k.\overline{T(S_1)}^o = k.\overline{T(S_1)}^o$ (see Exercise **89**).

Since Y is a Baire space, there is $k \geq 1$ such that $\overline{kT(S_1)}^o \neq \emptyset$; but $\overline{kT(S_1)} = k.\overline{T(S_1)}$, so $\overline{kT(S_1)}^o = k.\overline{T(S_1)}^o = k.\overline{T(S_1)}^o \neq \emptyset$; consequently there is $p \in \overline{T(S_1)}$ and $\eta > 0$ such that the open ball $B(p, \eta)$ is contained in $\overline{T(S_1)}$. Since $B(p, \eta) = p + B(0, \eta)$ we get $B(0, \eta) \subset \overline{T(S_1)} - p$; as $p \in \overline{T(S_1)}$, this gives $B(0, \eta) \subset \overline{T(S_1)} - \overline{T(S_1)} \subset \overline{T(S_0)}$. On the other hand it is clear that

$\overline{T(S_0)} = 2^n \overline{T(S_n)}$, so we deduce that $B\left(0, \dfrac{\eta}{2^n}\right) \subset \overline{T(S_n)}, \forall n \geq 0$.

Note that $S_0 = A$ and $T(S_0) = T(A)$. We prove now that $B\left(0, \dfrac{\eta}{2}\right) \subset T(A)$, and this will achieve the proof. Let $y \in B\left(0, \dfrac{\eta}{2}\right)$ then $y \in \overline{T(S_1)}$ and there is $x_1 \in S_1$ such that $\|y - T(x_1)\| < \dfrac{\eta}{4}$. This implies that $y - T(x_1) \in B\left(0, \dfrac{\eta}{4}\right) \subset \overline{T(S_2)}$, so there is $x_2 \in S_2$ such that $\|y - T(x_1) - T(x_2)\| < \dfrac{\eta}{8}$. If we continu the process in this way we get a sequence (x_n) in$X$ with $x_n \in S_n$ and

$$\left\| y - \sum_1^n T(x_j) \right\| < \frac{\eta}{2^{n+1}}, \forall n \geq 1.$$ Since $\|x_n\| < \dfrac{1}{2^n}$, the series $\sum_n x_n$ is absolutely convergent and so convergent because X is a Banach space. Put $x = \sum_n x_n$, then

$$\|x\| \leq \sum_n \|x_n\| < \sum_n \frac{1}{2^n} = 1,$$ this yields $x \in S_0 = A$ and, by the continuity of T,

$$T(x) = T\left(\sum_n x_n\right) = \sum_n T(x_n).$$ But from the nature of (x_n) we have $y = \sum_n T(x_n)$ and we conclude that $y = T(x) \in T(A)$. $\blacksquare$

Theorem 5.7.2.   (Open mapping theorem)
Let X,Y be Banach spaces and $T : X \longrightarrow Y$ a linear bounded operator. Suppose that T is onto, then T is an open mapping.

Proof: Let U be open in X, $x \in U$, and $y = T(x) \in T(U)$. There is $\delta > 0$ such that $B(x, \delta) \subset U$; since $B(x, \delta) = x + B(0, \delta)$ we get $B(0, \delta) \subset U - x$. Applying T which is linear, yields $T(B(0, \delta)) = \delta T(B(0, 1)) \subset T(U) - y$. By Lemma **5.7.1**, there is $\gamma > 0$ such that $B(0, \gamma) \subset T(B(0, 1))$; we get:
$\quad \delta B(0, \gamma) = B(0, \delta\gamma) \subset \delta T(B(0, 1)) \subset T(U) - y$.
So $y + B(0, \delta\gamma) = B(y, \delta\gamma) \subset T(U)$. This proves that $T(U)$ is open in Y. $\blacksquare$

Corollary: Let X,Y be Banach spaces and $T : X \longrightarrow Y$ a linear bijective operator. Suppose that T is bounded, then T^{-1} is bounded and so T is an isomorphism.

Proof: Let U be open in X. We have $\left(T^{-1}\right)^{-1}(U) = T(U)$. By the open mapping theorem $T(U)$ is open in Y, then T^{-1} is bounded. $\blacksquare$

Proposition 5.7.3. Let X be a vector space endowed with two norms $\|\cdot\|_1 , \|\cdot\|_2$ such that X is a Banach space for each of them. Suppose there is $\alpha > 0$ such that $\|x\|_2 \leq \alpha \|x\|_1 , \forall x \in X$, then there is $\beta > 0$ satisfying $\|x\|_1 \leq \beta \|x\|_2 , \forall x \in X$. In other words, two comparable norms on a Banach space are equivalent.

Proof: The condition says that that the identity map $i : X, \|\cdot\|_1 \longrightarrow X, \|\cdot\|_2$ is continuous, so by the corollary the inverse $i^{-1} : X, \|\cdot\|_2 \longrightarrow X, \|\cdot\|_1$ is continuous, whence the existence of $\beta > 0$ such that $\|x\|_1 \leq \beta \|x\|_2 , \forall x \in X$. $\blacksquare$

Definition 5.7.4. Let X,Y be normed spaces and $T : X \longrightarrow Y$ a linear operator. The graph of T is the subspace Γ of $X \times Y$ defined by $\Gamma = \{(x, T(x)) : x \in X\}$. We say that T is closed if its graph Γ is closed in the product space $X \times Y$.

Remark: Let (x_n) be a sequence in X and consider the conditions:
$\quad (i)\ x_n \longrightarrow x, n \longrightarrow \infty$
$\quad (ii)\ T(x_n) \longrightarrow y$
$\quad (iii)\ y = T(x)$
If T is closed then (i) and $(ii) \implies (iii)$. If T is continuous $(i) \implies (ii)$ and (iii).

Proposition 5.7.5. If $T : X \longrightarrow Y$ linear continuous then T is closed.

Proof: Let $\varphi : X \times Y \longrightarrow Y \times Y$ be the mapping defined by $\varphi(x,y) = (T(x),y)$. Then φ is continuous (why?) and we have $\Gamma = \varphi^{-1}(\Delta)$, where Δ is the diagonal of $Y \times Y$. Since Y is Hausdorff, Δ is closed and then Γ is closed.■

The converse of this proposition is false in general as is shown by the following:
Example: Consider the space $C[0,1]$ equipped with the uniform norm and the space $C^1[0,1]$ of functions $f : [0,1] \longrightarrow \mathbb{R}$ with first continuous derivative, also equipped with the uniform norm

Define $T : C^1[0,1] \longrightarrow C[0,1]$, by $T(x) = x' = \dfrac{dx}{dt}$.

T is not continuous since for example, if $x_n(t) = t^n, t \in [0,1], n \geq 1$, we have $\|x_n\| = 1$, but $\|T(x_n)\| = \sup_{0 \leq t \leq 1} n t^{n-1} = n \longrightarrow \infty$. However T is closed; indeed let $(x_n) \subset C^1[0,1]$ be such that $x_n \longrightarrow x$ and $T(x_n) \longrightarrow y$, then $x_n(t)$ converges uniformly to $x(t)$ and $T(x_n) = x'_n$ converges uniformly to y; consequently $x \in C^1[0,1]$ and $x' = y = T(x).$■

The remarkable fact is contained in the following:
Theorem 5.7.6.   (Closed graph theorem)
Let X,Y be Banach spaces and $T : X \longrightarrow Y$ a linear operator.
If T is closed then T is continuous.

Proof: The space X is a Banach space for the norm $\|\|x\|\| = \|x\| + \|T(x)\|$. Indeed let x_n be Cauchy for the norm $\|\|\cdot\|\|$; so we have $\|\|x_n - x_m\|\| = \|x_n - x_m\| + \|T(x_n) - T(x_m)\| \longrightarrow 0, n,m \longrightarrow \infty$. We get (x_n) Cauchy in X and $(T(x_n))$ in Y. Since X,Y are Banach spaces, the limit $x \in X$ of x_n and the limit $y \in Y$ of $T(x_n)$ and we have $y = T(x)$ because T is closed. This implies that $\|\|x_n - x\|\| = \|x_n - x\| + \|T(x_n) - T(x)\| \longrightarrow 0, n \longrightarrow \infty$ and proves that X is a Banach space for the norm $\|\|\cdot\|\|$. As $\|x\| \leq \|\|x\|\|$, there is $\gamma > 0$ such that $\|\|x\|\| \leq \gamma\|x\|$ (Proposition **5.7.3**); since $\|T(x)\| \leq \|\|x\|\|$, we deduce that $\|T(x)\| \leq \gamma\|x\|, \forall x \in X$ and T is continuous.■

Theorem 5.7.7. Let X be a Baire space (for example a metric complete space or a locally compact space). Let $(f_i, i \in I)$ be a family of continuous functions from X into $\mathbb{R}$ such that $\sup\{|f_i(x)|, i \in I\} < \infty$, for each $x \in X$. Then there is a nonempty open U of X and a constant $M > 0$ such that:
$$\sup\{|f_i(x)|, i \in I\} \leq M, \forall x \in U.$$

Proof: The condition on the f_i means that for each $x \in X$ there is $M_x > 0$ such that $|f_i(x)| \leq M_x, \forall i \in I$. Consider the set $A_{in} = \{x \in X : |f_i(x)| \leq n\}, i \in I, n \geq 1$; it is closed by the continuity of f_i, so the set $A_n = \bigcap_i A_{in}$ is also closed. On the other hand if $x \in X$ there is $n_x \geq 1$ such that $|f_i(x)| \leq n_x, \forall i \in I$, i.e $x \in A_{n_x}$. This proves that $X = \bigcup_n A_n$ and since X is a Baire space, there is $m \geq 1$ such that the interior set $\overset{o}{A}_m$ is non empty. Put $U = \overset{o}{A}_m$ and $M = m$, then we get $|f_i(x)| \leq M, \forall x \in U, \forall i \in I.$■

Theorem 5.7.8. (Uniform boundedness Theorem)
Let X be a Banach space and let $(E_i, i \in I)$ be a family of normed spaces.
For each $i \in I$ let $T_i : X \longrightarrow E_i$ a linear continuous operator such that:
$\sup \{\|T_i(x)\|, i \in I\} < \infty$, for each $x \in X$. Then we have: $\sup \{\|T_i\|, i \in I\} < \infty$.

Proof: For each $i \in I$ define the continuous function $f_i : X \longrightarrow \mathbb{R}$ by $f_i(x) = \|T_i(x)\|$. The condition reads: for each $x \in X$ there is $M_x > 0$ such that $f_i(x) = \|T_i(x)\| \leq M_x, \forall i \in I$. Since X is a Baire space there is a nonempty open U of X and a constant $M > 0$ such that $\sup \{\|T_i(x)\|, i \in I\} \leq M, \forall x \in U$ (Theorem **5.7.7**). Let $B(a, r)$ be an open ball contained in U with $a \in U$ and $r > 0$. So we have $\|x - a\| < r \implies \sup \{\|T_i(x)\|, i \in I\} \leq M$. We show that $\|T_i\| \leq \dfrac{2M}{r}, \forall i \in I$. To this end it is enough to have $\|T_i(y)\| \leq \dfrac{2M}{r}, \forall i \in I$ for $\|y\| < 1$. For such y put $x = a + ry$, we get $\|x - a\| = r\|y\| < r$ and then $\|T_i(x)\| \leq M, \forall i \in I$. This gives $\|T_i(a) + rT_i(y)\| \leq M, \forall i$. But $a \in U$, so $\|T_i(a)\| \leq M$.
Now we make the following estimation:
$\|rT_i(y)\| \leq \|T_i(a) + rT_i(y)\| + \|T_i(a)\| \leq M + M = 2M$ whence $\|T_i(y)\| \leq \dfrac{2M}{r}, \forall i \in I$. Finally we get $\|T_i\| = \sup \{\|T_i(y)\|, \|y\| < 1\} \leq \dfrac{2M}{r}, \forall i \in I.\blacksquare$

Corollary: (Banach-Steinhauss)
Let X be a Banach space and let E be a normed space. Let (T_n) be a sequence of linear bounded operators from X into E. Suppose that $T(x) = \lim\limits_{n} T_n(x)$ exists in E for each $x \in X$. Then $T(x)$ defines a linear bounded operator T from X into E with $\|T\| \leq \liminf\limits_{n} \|T_n\|$.

Proof: It is clear that T is linear. Let $\epsilon > 0$ and $x \in X$, there is $N = N_{\epsilon,x} \geq 1$ such that $\forall n \geq N : \|T_n(x)\| \leq \|T(x)\| + \epsilon$. Since $\sup\limits_{n \leq N} \|T_n(x)\| < \infty$, we deduce that $\sup\limits_{n} \|T_n(x)\| < \infty$, for each $x \in X$. By the uniform boundedness theorem, there is $M > 0$ such that $\sup\limits_{n} \|T_n\| \leq M$, this yields $\|T_n(x)\| \leq \|T_n\|\|x\| \leq M\|x\|, \forall n \geq 1$ and $\|T(x)\| = \lim\limits_{n} \|T_n(x)\| \leq M\|x\|$, this proves that T is bounded. On the other hand $\|T_n(x)\| \leq \|T_n\|\|x\|, \forall n \geq 1 \implies \|T(x)\| = \lim\limits_{n} \|T_n(x)\| \leq \liminf\limits_{n} \|T_n\|\|x\|$ and then $\|T\| \leq \liminf\limits_{n} \|T_n\|.\blacksquare$

8. Exercises

103. Let X, Y be Banach spaces and $T : X \longrightarrow Y$ a linear operator such that $g \circ T \in X^*$ for every $g \in Y^*$. Prove that T is bounded, (use the closed graph theorem).

104. Let X, Y be Banach spaces and $T : X \times Y \longrightarrow \mathbb{R}$ be a bilinear mapping, $X \times Y$ equipped with the product topology. Suppose that the partial functions:
$$T_x : Y \longrightarrow \mathbb{R}, \quad T_x(y) = T(x, y)$$
$$T_y : X \longrightarrow \mathbb{R}, \quad T_y(x) = T(x, y)$$
are continuous. The aim is to prove that T is continuous. It is enough to prove that T is continuous at $(0, 0)$ (Proposition **5.3.10**). Let (x_n, y_n) be a sequence in $X \times Y$ converging to $(0, 0)$. Define the linear operator $T_n : Y \longrightarrow \mathbb{R}, \quad T_n(y) = T(x_n, y)$.
(a) Prove that the set $\{|T_n(y)|, n \geq 1\}$ is bounded for each $y \in Y$.

(b) Prove that $M = \sup_n \|T_n\| < \infty$ and $|T_n(y)| \leq M \|y\|, \forall y, \forall n$.

(c) Deduce that $|T(x_n, y_n)| \longrightarrow 0$.

105. Let T be a linear bijection from a normed space X to a normed space Y. Prove that T is closed if and only if T^{-1} is closed.

106. Let X, Y be compact topological spaces and let us equip the spaces $C(X), C(Y)$ with the uniform norm. A linear operator $T : C(X) \longrightarrow C(Y)$ is positif if $f \geq 0 \implies T(f) \geq 0$.
(a) Prove that if T is positif then $f \geq g \implies T(f) \geq T(g)$.
(b) If T is positif then T is continuous and $\|T\| = \|T(1)\|$, where 1 is the constant function equal to 1.
(c) Let $T_n : C(X) \longrightarrow C(Y)$ be a sequence of linear operators such that $T_{n+1} - T_n$ is positif for every n and let $T : C(X) \longrightarrow C(Y)$ be a linear operator. Prove that $T_n \longrightarrow T$ for the operators norm iff $T_n(1) \longrightarrow T(1)$ for the uniform norm.

107. Let X, Y be normed spaces, $T : X \longrightarrow Y$ a linear operator and Γ the graph of T. Consider the closure $\overline{\Gamma}$ of Γ in $X \times Y$.
(a) Prove that $\overline{\Gamma}$ is the graph of a linear operator from X into Y if and only if
$$\forall y \in Y \setminus \{0\}, (0, y) \notin \overline{\Gamma}.$$
(b) Let $X \subset C[0,1]$ be the subspace of those $f \in C[0,1]$ that have $f'(0)$ (derivative at 0). Define $T : X \longrightarrow C[0,1]$ by $(Tf)(x) = f'(0)$, that is, Tf is the constant function equal to $f'(0)$. Let f_n be the sequence of functions given by $f_n(x) = \dfrac{\sin nx}{n}, x \in [0,1]$. Then $(Tf_n)(x) = f'_n(0) = 1, \forall n$. If Γ is the graph of T prove that (f_n, Tf_n) converges to $(0, 1)$. Deduce that $\overline{\Gamma}$ does not define a linear operator.

9. Duality in normed spaces. Weak topologies

Let X be a normed space with dual X^* and bidual X^{**}. We know that X^* and X^{**} are Banach spaces. Fix $x \in X$ and define the linear functional x^{**} on X^* by: $x^{**} : X^* \longrightarrow \mathbb{K}, x^{**}(f) = f(x), f \in X^*$. Then we have $|x^{**}(f)| = |f(x)| \leq \|f\| \|x\|$, so x^{**} is continuous i.e $x^{**} \in X^{**}$. More precisely we have:

Theorem 5.9.1. The mapping $\varphi : X \longrightarrow X^{**}$ given by $\varphi(x) = x^{**}$ is a linear isometry from X into X^{**}.

Proof: It is easy to check that φ is linear. We show that $\|\varphi(x)\| = \|x\|, \forall x \in X$, that is φ is an isometry. If $f \in X^*$, we have $|\varphi(x)(f)| = |f(x)| \leq \|x\| \|f\|$, so $\|\varphi(x)\| \leq \|x\|$. On the other hand if $x \in X$, there is $f \in X^*$, such that $f(x) = \|x\|$ and $\|f\| = 1$ (corollary **2** of Hahn-Banach theorem); for such f we have $|\varphi(x)(f)| = |f(x)| = \|x\| \leq \|\varphi(x)\| \|f\| = \|\varphi(x)\|$ whence $\|\varphi(x)\| \geq \|x\|$, this yields $\|\varphi(x)\| = \|x\|$. $\blacksquare$

Definition 5.9.2. In the sequel we put $X_0 = \varphi(X)$, so X_0 is a subspace of the bidual X^{**} and $\overline{X_0}$ is a Banach subspace of X^{**}. We say that a normed space X is reflexive if $X_0 = X^{**}$.

Example 5.9.3. Let $(S, \mathcal{F}, \mu)$ be a measure space, (for details on examples $(a), (b)$ see any basic course on integration).
(a) If $1 \leq p < \infty$ let L_p be the space of functions $f : S \longrightarrow \mathbb{R}$ such that $|f|^p$ is integrable. Then $L_p^* = L_q$ where q is the conjugate of p that is q satisfies $pq = p + q$. So we deduce that L_p is reflexive.

(b) Let L_∞ be the space of functions $f : S \longrightarrow \mathbb{R}$ bounded μ–almost everywhere and define the norm $\|f\|_\infty = \inf \{\alpha > 0 : \mu\{s : \mu(f > \alpha) = 0\}\}$, then L_∞ is a Banach space and if μ is σ–finite we have $L_\infty^* = L_p$.

(c) The space l_1 is not reflexive since $l_1^* = l_\infty$ and $l_\infty^* \neq l_1$ (see Exercise **100**).

(d) The space c_0 is not reflexive since $c_0^* = l_1$ and $c_0^{**} = l_\infty$.

Proposition 5.9.4. Every normed space X of finite dimension is reflexive and we have $\dim X = \dim X^* = \dim X^{**}$, where $\dim$ denotes the dimension.

Proof: Let $x_1, x_2, ..., x_n$ be a basis of X and Define $f_j : X \longrightarrow \mathbb{R}$ by:

if $x = \sum_{i=1}^{n} \lambda_j.x_j \in X$, $f_j(x) = \lambda_j, 1 \leq j \leq n$. Let us point out that f_j is linear and, since X is of finite dimension, f_j is continuous (Exercise **101**); so $f_j \in X^*$.
We prove that the $f_j, 1 \leq j \leq n$ form a basis of X^*. Let $f \in X^*$, $x \in X$ and

put $f(x_j) = \alpha_j$, then we have $f(x) = f\left(\sum_{i=1}^{n} \lambda_j.x_j\right) = \sum_{i=1}^{n} \lambda_j.\alpha_j$; but $\lambda_j = f_j(x)$,

so $f(x) = \sum_{i=1}^{n} \alpha_j.f_j(x), \forall x \in X$. This proves that the f_j generate X^*. On the

other hand suppose that $\sum_{i=1}^{n} \lambda_j.f_j = 0$, then for each $k \leq n$, $\sum_{i=1}^{n} \lambda_j.f_j(x_k) = \lambda_k = 0$,

whence $\lambda_1 = \lambda_2 = ... = \lambda_n = 0$, and the f_j are independent.■

The following theorem gives an interesting description of the topology generated on a vector space by a family of linear functionals. For the definition of the topology generated by a family of functions, see Theorem **2.3.11**, Chapter **2**.

Theorem 5.9.5. Let X be a vector space and let $\{f_i, i \in I\}$ be a family of linear functionals $f_i : X \longrightarrow \mathbb{R}$ on X. Consider the family $\mathcal{F}$ of all finite intersections $\bigcap_{i \in J} f_i^{-1}(V)$, where $J \subset I$ is finite and V is runing over a base of neighborhoods of 0 in $\mathbb{R}$. Then $\mathcal{F}$ is a base of neighborhoods of 0 in X for the topology $\tau = \tau\{f_i, i \in I\}$ generated by the family $\{f_i, i \in I\}$.

Proof: Let U be an open neighborhood of 0 in X for the topology $\tau = \tau\{f_i, i \in I\}$, from Theorem **2.3.11**, Chapter **2**, there is a finite family $U_j, j \in J$ in $\mathbb{R}$ such that $0 \in \bigcap_{j \in J} f_j^{-1}(U_j) \subset U$. But $f_j(0) = 0$ so $0 \in U_j, \forall j \in J$ and $V_0 = \bigcap_{j \in J} U_j$ is a neighborhood of 0 in $\mathbb{R}$, and we have $\bigcap_{j \in J} f_j^{-1}(V_0) = \bigcap_{j \in J} f_j^{-1}(U_j)$. If V is a basic neighborhoods of 0 in $\mathbb{R}$ with $V \subset V_0$ we deduce that $0 \in \bigcap_{j \in J} f_j^{-1}(V) \subset U$.■

Choosing basic neighborhoods of 0 in $\mathbb{R}$, the sets $V = \{x \in \mathbb{R} : |x| < \epsilon, \epsilon > 0\}$ we obtain:

Corollary: Under the conditions of Theorem **5.9.5**, the family of sets
$$V_{\epsilon, J} = \{x \in X : |f_j(x)| < \epsilon, j \in J\}, \epsilon > 0, J \text{ finite}$$
is a base of neighborhoods of 0 in X for the topology $\tau\{f_i, i \in I\}$.

Theorem 5.9.6. Let X be a vector space and let $\{f_i, i \in I\}$ be a family of linear functionals $f_i : X \longrightarrow \mathbb{K}$ on X. Then the mappings:

$\varphi : X \times X \longrightarrow X, \varphi(x, y) = x + y$

$h : \mathbb{K} \times X \longrightarrow X, h(\lambda, x) = \lambda.x$

are continuous, if X is endowed with the topology $\tau\{f_i, i \in I\}$, and the spaces $X \times X, \mathbb{K} \times X$ equipped with the product topology.

Proof: Use Proposition **2.3.12** of Chapter 2.∎

Corollary: Under the conditions of Theorem **5.9.6**, the family of sets

$x + V_{\epsilon,J}, \epsilon > 0, J$ finite, where $V_{\epsilon,J} = \{x \in X : |f_j(x)| < \epsilon, j \in J\}$

is a base of neighborhoods of x in X for the topology $\tau\{f_i, i \in I\}$.

Proof: By Theorem **5.9.6** the translation $t \longrightarrow t - x$ and its inverse $t \longrightarrow t + x$ are homeomorphisms of X for the topology $\tau\{f_i, i \in I\}$. So if U is an open neighborhood of x in X, $U - x$ is an open neighborhood of 0. By the corollary of Theorem **5.9.5**, there is $\epsilon > 0$ and a finite $J \subset I$ such that $V_{\epsilon,J} \subset U - x$ and then $x + V_{\epsilon,J} \subset U$.∎

Definition 5.9.7. Let X, Y be vector spaces and let $B : X \times Y \longrightarrow \mathbb{K}$ be a bilinear functional. We say that B is nondegenerate if:

$B(x, y) = 0, \forall y \in Y \implies x = 0$

$B(x, y) = 0, \forall x \in X \implies y = 0$

Definition 5.9.8. We say that there is a duality between vector spaces X, Y if there exist a nondegenerate bilinear functional $B : X \times Y \longrightarrow \mathbb{K}$.

Examples 5.9.9. (a) Let X be a vector space and let $Y = X'$ be the vector space of all linear forms $X \longrightarrow \mathbb{K}$, that is X' is the algebraic dual of X. Then the formula $B(x, x') = x'(x), x \in X, x' \in X'$ defines a duality between X, X' called the canonical duality. To see that B is nondegenerate, the reader is refered to any basic algebra course.

(b) Let X be a normed space with dual X^*, the Banach space of all continuous linear forms $X \longrightarrow \mathbb{K}$, then the bilinear functional $B(x, f) = f(x), x \in X, f \in X^*$ defines a duality between X, X^*. Indeed B is nondegenerate since, by the Hahn-Banach, if $x \neq 0$ there $f \in X^*$ such that $f(x) \neq 0$ (see corollary 2 of theorem 5.3.8).

Remark 5.9.10. Let X, Y be vector spaces and let $B : X \times Y \longrightarrow \mathbb{K}$ be a nondegenerate bilinear functional defining a duality between them. For $x \in X, y \in Y$ let us put $B_x(y) = B(x, y)$ and $B^y(x) = B(x, y)$. Then B_x (resp. B^y) is a linear functional on Y (resp. on X) and we have:

Proposition 5.9.11. (a) The mapping $x \longrightarrow B_x$ is a linear injection from X into the space of linear forms on Y.

(b) The mapping $y \longrightarrow B^y$ is a linear injection from Y into the space of linear forms on X.

Proof: It is a consequence of the fact that $B(x, y)$ is a nondegenerate bilinear functional.∎

Definition 5.9.12. Let X, Y be vector spaces with duality given by a nondegenerate bilinear functional $B(x, y)$. Then we define:

(a) The topology $\sigma(X, Y)$ as the topology on X generated by the family of mappings $\{B^y, y \in Y\}$.

(b) The topology $\sigma(Y, X)$ as the topology on Y generated by the family of mappings $\{B_x, x \in X\}$.

The description of the topology generated by a family of linear forms is given in Theorems **5.9.5** and **5.9.6**.

Definition 5.9.13. (**Weak topologies**)

Let X be a normed space:

(a) The topology $\sigma(X, X^*)$ is called the weak topology on X induced by the bilinear functional $B(x, f) = f(x), x \in X, f \in X^*$; so $\sigma(X, X^*) = \tau(f, f \in X^*)$.

(b) The topology $\sigma(X^*, X)$ is called the weak* (weak star) topology on X^* induced by the bilinear functional $B(f, x) = f(x), x \in X, f \in X^*$;

so $\sigma(X^*, X) = \tau(B_x, x \in X)$, where $B_x : X^* \longrightarrow \mathbb{K}$ is defined by $B_x(f) = f(x)$.

Theorem 5.9.14. (a) The family of sets $V_{\epsilon,n} = \{x \in X : |f_i(x)| < \epsilon, i = 1, 2, ..., n\}$ where $\epsilon > 0, n \geq 1, f_i \in X^*, i = 1, 2, ..., n$ is a base of neighborhoods of 0 for the weak topology $\sigma(X, X^*)$.

(b) The family of sets $V_{\epsilon,n}^* = \{f \in X^* : |f_i(x)| < \epsilon, i = 1, 2, ..., n\}$ where $\epsilon > 0, n \geq 1, x_i \in X, i = 1, 2, ..., n$ is a base of neighborhoods of 0 in X^* for the weak$-*$ topology $\sigma(X^*, X)$.

Proof: Apply Theorems **5.9.5** and **5.9.6**.■

As an immediate consequence of this theorem we have:

Theorem 5.9.15. (a) The family of sets $x + V_{\epsilon,n}, \epsilon > 0, n \geq 1$ is a base of neighborhoods of $x \in X$ for the weak topology $\sigma(X, X^*)$.

(b) The family of sets $g + V_{\epsilon,n}^*, \epsilon > 0, n \geq 1$ is a base of neighborhoods of $g \in X^*$ for the weak$-*$ topology $\sigma(X^*, X)$.

Theorem 5.9.16. The weak topology $\sigma(X, X^*)$ is compatible with the vector structure of X. Moreover the topology $\sigma(X, X^*)$ is Hausdorff. The same conclusions are valid for $X^*, \sigma(X^*, X)$.

Proof: The compatibility of $\sigma(X, X^*)$ and $\sigma(X^*, X)$ comes from Theorem **5.9.6**. To prove the Hausdorff property, let us show first that if $x \in X, x \neq 0$ there is a neighborhood V_x of x and a neighborhood V_0 of 0, in $\sigma(X, X^*)$, such that $V_x \cap V_0 = \emptyset$. By the Hahn-Banach theorem there is $f \in X^*$ such that $f(x) \neq 0$. On the other hand, since the field $\mathbb{K}$ is Hausdorff there is $\epsilon > 0$ and $\delta > 0$ such that the open sets $U_\epsilon = \{\lambda \in \mathbb{K} : |\lambda - f(x)| < \epsilon\}$ and $W_\delta = \{\lambda \in \mathbb{K} : |\lambda| < \delta\}$ are disjoint. Put $V_x = f^{-1}(U_\epsilon), V_0 = f^{-1}(W_\delta)$ then $V_x \cap V_0 = \emptyset$ and, since f is $\sigma(X, X^*)$ continuous, V_x, V_0 are open for the topology $\sigma(X, X^*)$. Now let $x \neq y$, by the preceding fact, there is a neighborhood V_{x-y} of $x - y$ and a neighborhood V_0 of 0, in $\sigma(X, X^*)$, such that $V_{x-y} \cap V_0 = \emptyset$. The sets $A_x = y + V_{x-y}$ and $B_y = y + V_0$ are open neighborhoods in $\sigma(X, X^*)$ of x and y respectively, by the compatibility of $\sigma(X, X^*)$; moreover they satisfy $A_x \cap B_y = y + (V_{x-y} \cap V_0) = \emptyset$, this proves that $X, \sigma(X, X^*)$ is Hausdorff. The proof is same for $X^*, \sigma(X^*, X)$, and even simpler since it does not use Hahn-Banach theorem.■

In the sequel we need to describe the $\sigma(X, Y)$ $-$linear forms. Let us start with the following algebraic lemma

Lemma 5.9.17. Let $f, f_1, f_2, ..., f_n$, be linear forms on a vector space X such that f is not 0. Let $M = \{x \in X : f(x) = 0\}$ be the kernel of f, and let $M_i = \{x \in X : f_i(x) = 0\}$ be the kernel of f_i, $i = 1, 2, ..., n$. Suppose $\overset{n}{\underset{1}{\cap}} M_i \subset M$, then there are scalars $\lambda_1, \lambda_2, ..., \lambda_n$ such that $f = \lambda_1 f_1 + \lambda_2 f_2 + ... + \lambda_n f_n$.

Proof: See any basic algebra course.∎

Theorem 5.9.18. Let X, Y be vector spaces with duality given by a nondegenerate bilinear functional $B(x, y)$. A linear form f on X is $\sigma(X, Y)$ continuous if and only if $f = B^y$ for some $y \in Y$.

Proof: If $f = B^y$ for some $y \in Y$, then f is $\sigma(X, Y)$ continuous since $\sigma(X, Y)$ is generated by the family of mappings $\{B^y, y \in Y\}$ (Definition **5.9.12.**). Conversely suppose that f is $\sigma(X, Y)$ continuous, then there is an open set of the form $V = \{x \in X : |B^{y_j}(x)| < \epsilon, j = 1, 2, ..., n\}$ such that $V \subset \{x \in X : |f(x)| < 1\}$. Put $\lambda(x) = \max_j |B^{y_j}(x)|$; we have $\{x \in X : |\lambda(x)| < \epsilon\} = V$ and then

$\{x \in X : \epsilon^{-1}|\lambda(x)| < 1\} \subset \{x \in X : |f(x)| < 1\}$. Let $x \in X$ and $\alpha > 0$, applying λ to the vector $y = \epsilon(\lambda(x) + \alpha)^{-1} x$ we get $\lambda(y) = \epsilon(\lambda(x) + \alpha)^{-1} \lambda(x) < \epsilon$; by the above inclusion, this implies that $|f(y)| < 1$ which gives $|f(y)| = \epsilon(\lambda(x) + \alpha)^{-1}|f(x)| < 1$ or $|f(x)| < \epsilon^{-1}(\lambda(x) + \alpha), \forall \alpha > 0$. If $\alpha \longrightarrow 0$ we get $|f(x)| < \epsilon^{-1}\lambda(x), \forall x \in X$ and $f(x)$ must be 0 if $x \in \underset{j}{\cap}\{x \in X : B^{y_j}(x) = 0\}$. By

Lemma **5.9.17** there are scalars $\mu_1, \mu_2, ..., \mu_n$ such that
$$f = \mu_1 B^{y_1} + \mu_2 B^{y_2} + ... + \mu_n B^{y_n}$$
but then $f = B^y$, with $y = \mu_1 y_1 + \mu_2 y_2 + ... + \mu_n y_n$.∎

Theorem 5.9.19. Let X be a normed space with norm topology τ, then:
(a) $\sigma(X, X^*) \subset \tau$.
(b) $\sigma(X, X^*) = \tau$ if and only if X is finite dimensional.

Proof: (a) The topology τ makes continuous the linear forms $f \in X^*$ and $\sigma(X, X^*)$ is the smallest one having this property, so $\sigma(X, X^*) \subset \tau$.

(b) If X is finite dimensional, the norm topology is the only Hausdorff topology compatible with the vector structure of X (see Tychonoff Theorem Appendix 3). By Theorem **5.9.16** we deduce the equality $\sigma(X, X^*) = \tau$. Conversely suppose that $\sigma(X, X^*) = \tau$. Let $V = \{x \in X : \|x\| < 1\}$, then V is also a $\sigma(X, X^*)$ neighborhood of 0; so there is $\epsilon > 0, n \geq 1$, and $f_1, f_2, ..., f_n$ in X^* such that $\{x \in X : |f_i(x)| < \epsilon, i = 1, 2, ..., n\} \subset V$. On the other hand if $f_i(x) = 0, i = 1, 2, ..., n$, then $x = 0$; indeed $\forall k \geq 1$ we have $f_i(kx) = kf_i(x) = 0, i = 1, 2, ..., n$, this shows that $\forall k \geq 1 \ kx \in V$, i.e $\|x\| < \dfrac{1}{k}, \forall k \geq 1$ so $x = 0$. Consequently the intersection of the kernels of $f_1, f_2, ..., f_n$ is $\{0\}$; from Lemma **5.9.17** we deduce that $f = \lambda_1 f_1 + \lambda_2 f_2 + ... + \lambda_n f_n$, for some scalars $\lambda_1, \lambda_2, ..., \lambda_n$. This proves that X^* is finite dimensional, and then X also is finite dimensional.∎

Let us observe that on the space X^* we can define three topologies that are: the strong topology induced by the norm, the weak topology $\sigma(X^*, X^{**})$ induced by mean of the canonical duality between X^* and X^{**} and finally the weak* topology. Relations between these topologies and some fundamental facts are given below, for the proofs the reader is refered to [2]. We start with a definition:

Definition 5.9.20. Consider the image X_0 of X by the canonical isometry $\varphi : X \longrightarrow X^{**}, \varphi(x)(f) = f(x), x \in X, f \in X^*$. Now define the bilinear form $B : X^* \times X_0 \longrightarrow \mathbb{K}, B(f, \varphi(x)) = \varphi(x)(f) = f(x)$. The bilinear form B is non degenerate and defines $\sigma(X^*, X_0)$ and $\sigma(X_0, X^*)$.

Theorem 5.9.21. With the above notations we have:
(a) $\sigma(X^*, X_0) = \sigma(X^*, X)$.
(b) The isomorphism φ from X onto $X_0 = \varphi(X)$ is bicontinuous for the topologies $\sigma(X, X^*)$ on X and $\sigma(X_0, X^*)$ on X_0.
(c) The relative topology induced on X_0 by the weak$-*$ topology $\sigma(X^{**}, X^*)$ is $\sigma(X_0, X^*)$.
(d) The canonical isometry $\varphi : X \longrightarrow X^{**}$ is continuous for the topologies $\sigma(X, X^*)$ on X and $\sigma(X^{**}, X^*)$ on X^{**}.

Theorem 5.9.22. Let τ^* be the norm topology on X^* then:
(a) $\sigma(X^*, X) \subset \sigma(X^*, X^{**}) \subset \tau^*$.
(b) $\sigma(X^*, X) = \sigma(X^*, X^{**})$ if and only if X is reflexive.
(c) $\tau^* = \sigma(X^*, X^{**})$ if and only if X is finite dimensional.

Theorem 5.9.23. (Alaoglu)
The unit ball of X^*, that is $B = \{f \in X^* : \|f\| \leq 1\}$, is compact for the weak* topology $\sigma(X^*, X)$.

10. Exercises

108. Let X be a normed space, prove that:
$\|f\| = \sup\{|f(x)|, \|x\| \leq 1\}$, if $f \in X^*$.
$\|x\| = \sup\{|f(x)|, f \in X^*, \|f\| \leq 1\}$, if $x \in X$.

109. (a) Let (x_n) be a sequence in X. Prove that x_n converges to x for the weak topology if and only if for every $f \in X^*$, the sequence $f(x_n)$ converges to $f(x)$ in $\mathbb{K}$. We say that x_n converges weakly to x.
(b) Prove that the convergence of functions for the weak$-*$ topology is identical to the simple convergence.

109. Prove that a sequence x_n converges weakly (**109**(a)) to x if and only if:
(1) $\sup_n \|x_n\| < \infty$
(2) $\lim_n f(x_n) = f(x), \forall f \in X^*$

110. (a) The norm convergence implies the weak convergence.
(b) The converse of (a) is not true: take the sequence $x_n = (\delta_{mn})_m$ in the space c_0 endowed with the uniform norm and use the fact that $c_0^* = l_1$ (Exercise **100**(c)).

111. In the Banach space l_1 the norm convergence is equivalent to the weak convergence (Apply V.H.S Theorem Appendix 4).

112. Let X_0 be the image of X by the canonical isometry $\varphi : X \longrightarrow X^{**}$ (Definition **5.9.2**). Prove that X_0 is dense in X^{**} for the weak$-*$ topology $\sigma(X^{**}, X^*)$.

113. A subset A of a normed space X is bounded if $\sup\{\|x\|, x \in A\} < \infty$; it is weakly bounded if $\sup\{|f(x)|, x \in A\} < \infty, \forall f \in X^*$. In a Banach space A is bounded if and only if it is weakly bounded.

114. Let X be a normed space with norm topology τ, $M \subset X$ a vector subspace. Prove that M is $\tau-$closed if and only if it is $\sigma(X, X^*)-$closed.

115. In a reflexive Banach space, prove that the unit ball $B = \{x \in X : \|x\| \leq 1\}$ is $\sigma(X, X^*)-$compact.[use the canonical isometryof X into X^{**} and apply Alaoglu theorem].

116. Let X, Y be normed spaces and $T : X \longrightarrow Y$ a linear continuous operator. We define the operator $T^* : Y^* \longrightarrow X^*$ by: $g \in Y^*, T^*(g) = g \circ T$.
(a) Prove that T^* is linear continuous (T^* is the adjoint of T).
(b) Let $S : X \longrightarrow Y$ and $T : Y \longrightarrow Z$ be linear continuous.Prove that
   $(T \circ S)^* = S^* \circ T^*$.
(c) Let $T^{**} : X^{**} \longrightarrow Y^{**}$ the adjoint of T^*. Prove that $T^{**} \circ \varphi(x) = \psi \circ T(x)$, $x \in X$ where φ is the canonical isometryof X into X^{**} and ψ the canonical isometryof Y into Y^{**}. Deduce from (a) that $\|T\| = \|T^*\|$.

CHAPTER 6

Hilbert Spaces

Hilbert spaces are a natural generalization of the euclidean spaces of finite dimension $\mathbb{R}^n$, which means that several familiar geometric properties of $\mathbb{R}^n$ can be extended to infinite dimensional spaces. Let us recall that the eucliden norm of $\mathbb{R}^n$:

$$x = (x_1, x_2, ..., x_n) \in \mathbb{R}^n, \ \|x\| = \left(\sum_{i=1}^{n} |x_i|^2 \right)^{\frac{1}{2}}$$

is induced by the inner product:

$$x = (x_1, x_2, ..., x_n), y = (y_1, y_2, ..., y_n), \quad \langle x, y \rangle = \sum_{i=1}^{n} x_i \cdot y_i$$

which is a bilinear form on $\mathbb{R}^n$ satisfying $\|x\| = \langle x, x \rangle^{\frac{1}{2}}$.

1. Hermitian Forms

In what follows X is a vector space on the scalar field $\mathbb{K} = \mathbb{R}$ or $\mathbb{C}$, whose nul vector will be denoted by 0.

Definition 6.1.1. An hermitian form on X is a function $\varphi : X \times X \longrightarrow \mathbb{K}$ satisfying the conditions:

(i) For each fixed y in X the function $x \longrightarrow \varphi(x, y)$ from X into $\mathbb{K}$ is linear.

(ii) (Hermitian symmetry) For all x, y in X we have $\overline{\varphi(x, y)} = \varphi(y, x)$, where $\overline{z}$ is the conjugate of z.

Remark 6.1.2.

(a) The hermitian symmetry implies that $\varphi(x, x)$ is real for all x in X.

(b) The hermitian symmetry implies also that:
$$\varphi(x, y + z) = \overline{\varphi(y + z, x)} = \overline{\varphi(y, x) + \varphi(z, x)} = \overline{\varphi(y, x)} + \overline{\varphi(z, x)}$$
$$= \varphi(x, y) + \varphi(x, z).$$
$$\varphi(x, \lambda y) = \overline{\varphi(\lambda y, x)} = \overline{\lambda \varphi(y, x)} = \overline{\lambda} \overline{\varphi(y, x)} = \overline{\lambda} \varphi(x, y),$$
for all x, y in X and all $\lambda \in \mathbb{K}$.

Definition 6.1.3. An hermitian form φ on X is said to be non degenerated if:
$$\varphi(x, x) \geq 0, \text{ for all } x \in X \text{ and } \varphi(x, x) = 0 \iff x = 0$$
An inner product on X is a non degenerated hermitian form on X and will be denoted by: $\langle x, y \rangle, \ x, y \in X$.

Theorem 6.1.4. If $\langle , \rangle$ is an inner product on X, then:

(a) (Cauchy-Schwarz inequality) For every $x, y \in X$ we have:
$$|\langle x, y \rangle|^2 \leq \langle x, x \rangle . \langle y, y \rangle$$

(b) The formula: $\|x\| = \langle x, x \rangle^{\frac{1}{2}}$, $x \in X$, defines a norm on X.

Proof: (a) For every $x, y \in X$ and $\lambda \in \mathbb{K} = \mathbb{R}$ or $\mathbb{C}$, we have $\langle x + \lambda y, x + \lambda y \rangle \geq 0$. Developping this inner product, we get:
$$\langle x + \lambda y, x + \lambda y \rangle = \langle x, x \rangle + \lambda \langle y, x \rangle + \overline{\lambda} \langle x, y \rangle + \lambda \overline{\lambda} \langle y, y \rangle$$
if $\langle y, y \rangle = 0$ then $y = 0$ and $\langle x, y \rangle = 0$, so (a) is satisfied in this case.

if $\langle y, y \rangle \neq 0$ taking $\lambda = -\dfrac{\langle x, y \rangle}{\langle y, y \rangle}$, we get:
$$\langle x + \lambda y, x + \lambda y \rangle = \langle x, x \rangle - \frac{|\langle x, y \rangle|^2}{\langle y, y \rangle} - \frac{|\langle x, y \rangle|^2}{\langle y, y \rangle} + \frac{|\langle x, y \rangle|^2}{\langle y, y \rangle} = \langle x, x \rangle - \frac{|\langle x, y \rangle|^2}{\langle y, y \rangle} \geq 0$$
therefore $|\langle x, y \rangle|^2 \leq \langle x, x \rangle . \langle y, y \rangle$, and (a) is satisfied.

(b) If $\|x\| = 0$ then $\langle x, x \rangle = 0$ and $x = 0$. On the other hand for $\lambda \in \mathbb{K}$ and $x \in X$, we have $\|\lambda x\| = \left(\lambda \overline{\lambda} \langle x, x \rangle \right)^{\frac{1}{2}} = |\lambda| \, \|x\|$.

To see the triangle inequality, we have:
$$\|x + y\|^2 = \langle x + y, x + y \rangle = \|x\|^2 + \|y\|^2 + \langle x, y \rangle + \langle y, x \rangle$$
$$= \|x\|^2 + \|y\|^2 + 2 \operatorname{Re}(\langle x, y \rangle) \leq \|x\|^2 + \|y\|^2 + 2 |\langle x, y \rangle| \leq \|x\|^2 + \|y\|^2 + 2 \|x\|^2 \|y\|^2$$
$$= (\|x\| + \|y\|)^2, \text{ so } \|x + y\| \leq \|x\| + \|y\| . \blacksquare$$

Definition 6.1.5. A prehilbert space $X, \langle . \rangle$ is a vector space endowed with an inner product $\langle . \rangle$.

We say that $X, \langle . \rangle$ is a Hilbert space if it is complete with respect to the norm induced by the inner product $\langle . \rangle$, as given in the theorem above.

Examples 6.1.6. (a) Let $X = C[0, 1] = \{f : [0, 1] \longrightarrow \mathbb{K}, \ f \text{ continuous}\}$.

For $f, g \in X$ define $\langle f, g \rangle = \int_0^1 f(t) . \overline{g(t)} dt$, the integral being a Riemann one. It is well known, from the Riemann integral properties, that X is not complete for the norm $\|f\| = \left(\int_0^1 |f(t)|^2 \, dt \right)^{\frac{1}{2}}$.

So X is a prehilbert space.

(b) Consider the space $l_2 = \left\{ (x_n) \in \mathbb{C}^{\mathbb{N}} : \sum_n |x_n|^2 < \infty \right\}$ and define:

For $x = (x_n)$, $y = (y_n)$ in l_2, $\langle x, y \rangle = \sum_n x_n . \overline{y_n}$ and $\|x\|_2 = \langle x, x \rangle^{\frac{1}{2}} = \left(\sum_n |x_n|^2 \right)^{\frac{1}{2}}$.

Then l_2 is a Hilbert space for the norm $\|x\|_2$, see example **5.1.14** (2) in chapter 5.

Proposition 6.1.7. (Parallelogram law)

In a prehilbert space X we have:
$$\|x + y\|^2 + \|x - y\|^2 = 2 \left(\|x\|^2 + \|y\|^2 \right), \forall x, y \in X.$$

Proof: We have:
$$\|x + y\|^2 = \langle x + y, x + y \rangle = \|x\|^2 + \|y\|^2 + 2 \operatorname{Re}(\langle x, y \rangle)$$
$$\|x - y\|^2 = \langle x - y, x - y \rangle = \|x\|^2 + \|y\|^2 - 2 \operatorname{Re}(\langle x, y \rangle)$$
we get the proof by addition. $\blacksquare$

2. Orthogonality

Definition 6.2.1. Let X be a prehilbert space. Two non zero distinct vectors x, y are orthogonal if $\langle x, y \rangle = 0$.

A family B of vectors in X is said to be orthogonal if for every $x \neq y$ in B we have $\langle x, y \rangle = 0$. B is orthonormal if it is orthogonal and $\|x\| = 1$, for all x in B.

Proposition 6.2.2. (Pythagorean theorem)

If the vectors x, y are orthogonal then $\|x + y\|^2 = \|x\|^2 + \|y\|^2$.

More generally, if $x_1, x_2, ..., x_n$ is a finite family of orthogonal vectors, we have:

$$\left\| \sum_1^n x_i \right\|^2 = \sum_1^n \|x_i\|^2 .$$

Proof: Since $\langle x, y \rangle = 0$, we have:

$$\|x + y\|^2 = \langle x + y, x + y \rangle = \|x\|^2 + \|y\|^2 + 2 \operatorname{Re}\left(\langle x, y \rangle \right) = \|x\|^2 + \|y\|^2 .$$

$$\left\| \sum_1^n x_i \right\| = \sum_1^n \|x_i\|^2 + \sum_{i \neq j} \langle x_i, x_j \rangle = \sum_1^n \|x_i\|^2, \text{ since } \langle x_i, x_j \rangle = 0, \forall i \neq j. \blacksquare$$

The proof of the following lemma uses the definition of the inner product and is left as exercise.

Lemma 6.2.3. Let $x_1, x_2, ..., x_n$ be a finite family of orthonormal vectors in a a prehilbert space X. Then for every $x \in X$ and all $a_1, a_2, ..., a_n$ in $\mathbb{K}$, we have:

$$\left\| x - \sum_1^n a_i x_i \right\|^2 = \|x\|^2 - \sum_1^n |\langle x, x_i \rangle|^2 + \sum_1^n |a_i - \langle x, x_i \rangle|^2 .$$

In many applications one aims to determine the best approximation of a vector $x \in X$ by a linear combination $\sum_1^n a_i x_i$ of some given vectors $x_1, x_2, ..., x_n$, in the sense of making the norm $\left\| x - \sum_1^n a_i x_i \right\|$ minimum. The following theorem gives the best way to do this.

Theorem 6.2.4. Let $x_1, x_2, ..., x_n$ be a finite orthonormal system of vectors in a prehilbert space X and let $x \in X$. Then the norm $\left\| x - \sum_1^n a_i x_i \right\|$, when the scalars are varying in $\mathbb{K}$, is minimal for $a_i = \langle x, x_i \rangle$, $i = 1, 2, ..., n$, and in this case the minimum is given by $\left(\|x\|^2 - \sum_1^n |\langle x, x_i \rangle|^2 \right)^{\frac{1}{2}}$. So we deduce that

$$\left\| x - \sum_1^n \langle x, x_i \rangle . x_i \right\| = \left(\|x\|^2 - \sum_1^n |\langle x, x_i \rangle|^2 \right)^{\frac{1}{2}} .$$

Proof: By Lemma **6.2.3** we have $\left\| x - \sum_1^n a_i x_i \right\|^2 = \|x\|^2 - \sum_1^n |\langle x, x_i \rangle|^2 + \sum_1^n |a_i - \langle x, x_i \rangle|^2 .$

Since $\left\| x - \sum_1^n a_i x_i \right\|^2$ is positive, the R.H.S is minimal iff $\sum_1^n |a_i - \langle x, x_i \rangle|^2 = 0$, that is $a_i = \langle x, x_i \rangle$, $i = 1, 2, ..., n$. In this case it is clear that the minimum value of $\left\| x - \sum_1^n a_i x_i \right\|$ is $\left(\|x\|^2 - \sum_1^n |\langle x, x_i \rangle|^2 \right)^{\frac{1}{2}} . \blacksquare$

As an application we give an important extension to infinite orthonormal sequences (x_n) in a prehilbert space:

Theorem 6.2.5. (Bessel Inequality)

Let (x_n) be an infinite orthonormal sequence of vectors in a prehilbert space X and let $x \in X$. Then we have $\|x\|^2 \geq \sum_n |\langle x, x_n \rangle|^2$, where the equality occurs iff

$$\left\| x - \sum_1^n \langle x, x_i \rangle . x_i \right\| \longrightarrow 0, n \longrightarrow \infty.$$

Proof: For each $n \geq 1$ and each finite subsequence $x_1, x_2, ..., x_n$ we have:

$$0 \leq \left\| x - \sum_1^n \langle x, x_i \rangle . x_i \right\|^2 = \|x\|^2 - \sum_1^n |\langle x, x_i \rangle|^2$$

this comes from Theorem **6.2.4**, and gives $\|x\|^2 \geq \sum_1^n |\langle x, x_i \rangle|^2$; letting $n \longrightarrow \infty$, we get the inequality.

Finally it is clear that equality occurs iff $\left\| x - \sum_1^n \langle x, x_i \rangle x_i \right\| \longrightarrow 0, n \longrightarrow \infty.\blacksquare$

Let M be a subset of a prehilbert space X and let $x \in X \backslash M$. Let us ask the following problem: does there exist a vector $y \in M$ giving the best approximation of x, that is, such that the norm $\|x - y\|$ is minimal. If X is a Hilbert space and M is a closed convex subset of X, the solution is given by:

Theorem 6.2.5. Let M be closed convex subset in a Hilbert space X. Then for each $x \in X$, there is a unique vector $y_0 \in M$ such that:

$$\|x - y_0\| = \inf \left\{ \|x - y\| : y \in M \right\}.$$

Proof: Let us recall that M convex means
$$\lambda \in \mathbb{K}, 0 \leq \lambda \leq 1 \implies \lambda M + (1 - \lambda) M \subset M.$$
Put $d = \inf \left\{ \|x - y\| : y \in M \right\}$, so there is a sequence (y_n) in M such that
$$\|x - y_n\| \longrightarrow d, n \longrightarrow \infty.$$
We show that (y_n) is a Cauchy sequence in X. By the Parallelogram law, we have:
$$\|(y_n - x) + (y_m - x)\|^2 + \|(y_n - x) - (y_m - x)\|^2 = 2 \left(\|y_n - x\|^2 + \|y_m - x\|^2 \right)$$
which gives:
$$\|y_n - y_m\|^2 = 2 \left(\|y_n - x\|^2 + \|y_m - x\|^2 \right) - 4 \left\| \tfrac{1}{2} (y_n + y_m) - x \right\|^2$$
since M is convex $\tfrac{1}{2} (y_n + y_m) \in M$, and $\left\| \tfrac{1}{2} (y_n + y_m) - x \right\| \geq d$
then we deduce that
$$\|y_n - y_m\|^2 \leq 2 \left(\|y_n - x\|^2 + \|y_m - x\|^2 \right) - 4d^2 \longrightarrow 0, \quad n, m \longrightarrow \infty.$$
As X is complete, y_n converges to some $y_0 \in M$, because M is closed. So we deduce that $\|x - y_n\| \longrightarrow \|x - y_0\| = d$, by the continuity of the norm.
Now we prove uniqueness. Take $z_0 \in M$ with $\|x - z_0\| = \|x - y_0\| = d$ and apply the Parallelogram law to get:
$$\|(y_0 - x) + (z_0 - x)\|^2 + \|(y_0 - x) - (z_0 - x)\|^2 = 2 \left(\|y_0 - x\|^2 + \|z_0 - x\|^2 \right) = 4d^2$$
So $\|y_0 + z_0 - 2x\|^2 + \|y_0 - z_0\|^2 = 2 \left(\|y_0 - x\|^2 + \|z_0 - x\|^2 \right) = 4d^2$
Since $\left\| y_0 + z_0 - 2x^2 \right\| = 4 \left\| \tfrac{1}{2} (y_0 + z_0) - x \right\|^2 \geq 4d^2$ we get $\|y_0 - z_0\| \leq 0$
that is $y_0 = z_0.\blacksquare$

Theorem 6.2.6. Let M be closed subspace of a Hilbert space X and let $x \in X \backslash M$. Then we have:

$y_0 \in M$ and $\|x - y_0\| = \inf\{\|x - y\| : y \in M\} \iff y_0 \in M$ and $x - y_0 \perp M$

where $x - y_0 \perp M$ means $\langle x - y_0, y \rangle = 0, \forall y \in M$.

Proof: Suppose $y_0 \in M$ and $x - y_0 \perp M$. Then:

$\forall y \in M, \|x - y\|^2 = \|(x - y_0) - (y - y_0)\|^2 = \|x - y_0\|^2 + \|y - y_0\|^2$

by Pythagorean theorem, because $x - y_0 \perp M$ and $y - y_0 \in M$. Consequently $\|x - y\|^2 \geq \|x - y_0\|^2$. So we get $\|x - y_0\| = \inf\{\|x - y\| : y \in M\}$.

Conversely suppose $\|x - y_0\| = \inf\{\|x - y\| : y \in M\}$. Let $y \in M$ and $c \in \mathbb{R}$, so $y_0 + cy \in M$ and then $\|x - y_0 - cy\| \geq \|x - y_0\|$. On the other hand we have:

$\|x - y_0 - cy\|^2 = \|x - y_0\|^2 + |c|^2 \|y\|^2 - 2\operatorname{Re}\langle x - y_0, cy \rangle$

we deduce that $|c|^2 \|y\|^2 - 2\operatorname{Re}\langle x - y_0, cy \rangle = \|x - y_0 - cy\|^2 - \|x - y_0\|^2 \geq 0$.

Taking $c = \alpha \langle x - y_0, y \rangle$, where α is real we get:

$\langle x - y_0, cy \rangle = \alpha |\langle x - y_0, y \rangle|^2$ and $\operatorname{Re}\langle x - y_0, cy \rangle = \alpha |\langle x - y_0, y \rangle|^2$

Then the above inequality gives $|\langle x - y_0, y \rangle|^2 \left(\alpha^2 \|y\|^2 - 2\alpha\right) \geq 0$. But with the choice $0 < \alpha \leq \dfrac{2}{\|y\|}$, we have $\alpha^2 \|y\|^2 - 2\alpha \leq 0$, so we deduce $\langle x - y_0, y \rangle = 0$. ∎

Corollary 1. Let M be closed subspace of a Hilbert space X. Then every vector $x \in X$ has a unique representation of the form:

$x = y + z$, with $y \in M$ and $z \perp M$

in fact y is the orthogonal projection of the vector x on the subspace M.

Proof: Let y_0 be the orthogonal projection of the vector x on M. Then the vectors $y = y_0$ and $z = x - y$ give the desired representation since by the preceding theorem $x - y_0 \perp M$. Let us prove the uniqueness. Suppose $x = y + z = y' + z'$ with $y, y' \in M, z, z' \perp M$. But $y - y' \in M$ and $y - y' = z - z' \perp M$, so $\langle y - y', y - y' \rangle = 0$, thus $y - y' = 0 = z - z'$. ∎

Corollary 2. If M is a subset of M, put $M^\perp = \{x \in X, \ x \perp M\}$. Then $M^\perp$ is a closed subspace of X. Moreover if M is a closed subspace of X, we have

$X = M \oplus M^\perp$, that is, X is the direct sum of M and $M^\perp$.

Proof: observe that $M^\perp$ is a closed subspace of X even if M is not a subspace and apply corollary 1. ∎

Definition 6.2.7. The vector y_0 satisfying $\langle x - y_0, y \rangle = 0, \forall y \in M$, is called the orthogonal projection of the vector x on the subspace M.

3. Orthonormal Bases

In this section we will have to deal with arbitrary families (x_α) of vectors in a prehilbert space and we should be able to define sums like $\sum_\alpha \langle x, x_\alpha \rangle x_\alpha$. Let us start with the summability of arbitrary families of scalars.

Definition 6.3.1. Let $\{x_\alpha, \alpha \in I\}$ be an arbitrary family of positive real numbers. We say that $\{x_\alpha, \alpha \in I\}$ is summable if the family of finite sums $\left\{\sum_{\alpha \in F} x_\alpha, F \subset I, \ F \text{ finite}\right\}$ is bounded. In this case we put $\sum_\alpha x_\alpha = \sup\left\{\sum_{\alpha \in F} x_\alpha, F \subset I, \ F \text{ finite}\right\}$.

Theorem 6.3.2. Let $\{x_\alpha, \alpha \in I\}$ be a summable family of positive real numbers. Then the set $D = \{\alpha \in I, \; x_\alpha \neq 0\} = \{\alpha \in I, \; x_\alpha > 0\}$ is at most countable.

Proof: For each integer $n \geq 1$, put $D_n = \left\{\alpha \in I, \; x_\alpha > \dfrac{1}{n}\right\}$.

Then $\{x_\alpha, \alpha \in I\}$ summable implies D_n finite for all n, otherwise

$\left\{\displaystyle\sum_{\alpha \in F} x_\alpha, F \subset I, \; F \text{ finite}\right\}$ would not be bounded, contradicting summability.

Since $D = \underset{n}{\cup} D_n$, we deduce that D is at most countable. $\blacksquare$

Remark 6.3.3. By this theorem, if $\{x_\alpha, \alpha \in I\}$ is summable, there is a sequence $(\alpha_n) \subset I$ such that $\forall \alpha \in I \backslash (\alpha_n)$, $x_\alpha = 0$ and we have $\displaystyle\sum_\alpha x_\alpha = \sum_n x_{\alpha_n}$.

Remark 6.3.4. The remark above allows to formulate Bessel inequality (Theorem **6.2.5**) as:

Let $B = \{x_\alpha, \alpha \in I\}$ be an orthonormal family of vectors in a prehilbert space X. Then $\forall x \in X$ there is a sequence $(\alpha_n) \subset I$ such that $\forall \alpha \in I \backslash (\alpha_n)$, $\langle x, x_\alpha \rangle = 0$ and $\|x\|^2 \geq \displaystyle\sum_n |\langle x, x_{\alpha_n} \rangle|^2$.

Definition 6.3.5. Let $\{x_\alpha, \alpha \in I\}$ be an arbitrary family of real or complex numbers. We say that $\{x_\alpha, \alpha \in I\}$ is absolutely summable if the family of moduli $\{|x_\alpha|, \alpha \in I\}$ is summable in the sense of Definition **6.3.1.**

In this case there is a sequence $(\alpha_n) \subset I$ such that $\forall \alpha \in I \backslash (\alpha_n)$, $|x_\alpha| = x_\alpha = 0$, $\displaystyle\sum_\alpha |x_\alpha| = \sum_n |x_{\alpha_n}|$. So we define the sum of the family $\{x_\alpha, \alpha \in I\}$ by $\displaystyle\sum_\alpha x_\alpha = \sum_n x_{\alpha_n}$.

Definition 6.3.6. Let $B = \{x_\alpha, \alpha \in I\}$ be an orthonormal family of vectors in a prehilbert space X. We say that B is maximal if any orthonormal family C containing B coincides with B. We say that B is an orthonormal base for the prehilbert space X, if B is maximal.

Let us recall that the subspace generated by a subset $B \subset X$ is defined by

$$M(B) = \left\{x \in X : x = \sum_1^n a_i x_i, n \geq 1, a_i \in \mathbb{K}, x_i \in B\right\}$$

(finite linear combinations of vectors in B)

The closure $\overline{M(B)}$ of the subspace $M(B)$ is also a subspace of X. In fact $\overline{M(B)}$ is the closed subspace generated by a subset B, see Chapter 5 for details.

For a Hilbert space, the following theorem gives an exhaustive characterization of orthonormal bases:

Theorem 6.3.7. Let $B = \{x_\alpha, \alpha \in I\}$ be an orthonormal family of vectors in a Hilbert space X. Then the following properties are equivalent:

    (a) B is an orthonormal base for the space X

    (b) If $x \in X$ and $x \perp B$ then $x = 0$

    (c) $\overline{M(B)} = X$

    (d) For all $x \in X$: $x = \displaystyle\sum_\alpha \langle x, x_\alpha \rangle . x_\alpha$

    (e) For all $x, y \in X$: $\langle x, y \rangle = \displaystyle\sum_\alpha \langle x, x_\alpha \rangle . \langle x_\alpha, y \rangle$

    (f) For all $x \in X$: $\|x\|^2 = \displaystyle\sum_\alpha |\langle x, x_\alpha \rangle|^2$ (**Parseval Relation**)

The numbers $\langle x, x_\alpha \rangle$ are called the Fourier coeffients of the vector x with respect to the orthonormal base B.

Proof: See [2] .

Theorem 6.3.8. Every non trivial Hilbert space has an orthonormal base.

Proof: Let $\mathfrak{S}$ be the set of all orthonormal families in a given Hilbert space and put on $\mathfrak{S}$ the inclusion ordering. We prove that every chain F in $\mathfrak{S}$ has an upper bound in $\mathfrak{S}$. To this end let $S = \underset{s \in F}{\cup} s$, then it is clear that S is an upper bound for F; moreover S is orthonormal. Indeed, let $x_1, x_2 \in S$ and $s_1, s_2 \in F$, with $x_1 \in s_1, x_2 \in s_2$; since F is a chain (that is totally ordered family) we can assume that $s_1 \subset s_2$. So both x_1, x_2 are in s_2, but s_2 is orthonormal, therefore $\|x_1\| = \|x_2\| = 1$ and $x_1 \perp x_2$. We deduce that S is orthonormal. Now apply Zorn lemma to get that $\mathfrak{S}$ has a maximal element B which is an orhnormal base for the given Hilbert space.■

The reader is refeered to Chapter 1 for all concepts attached to partial ordering.

The following definition gives an important example of Hilbert space.

Definition 6.3.9. Let A be an arbitrary non empty set and consider the following:

$$l_2(A) = \left\{ (x_\alpha)_{\alpha \in A} \subset \mathbb{K} : \sum_\alpha |x_\alpha|^2 < \infty \right\}$$

Note that $x = (x_\alpha)_{\alpha \in A} \in l_2(A)$, implies, by Theorem **6.3.2**, the existence of a countable set $D = D_x$ such that $x_\alpha = 0, \forall \alpha \in A \backslash D_x$. With these ingredients,

$$l_2(A) \text{ is a vector space and } \|x\|_2 = \left(\sum_\alpha |x_\alpha|^2 \right)^{\frac{1}{2}} \text{ is a norm making it a Banach}$$

space (The proof of these facts is similar to that given in Example **5.1.14** (2), Chapter 5).

Now let $x = (x_\alpha), y = (y_\alpha)$ be in $l_2(A)$, so there are countable sets $D_x, D_y \subset A$ such that $x_\alpha = 0, \forall \alpha \in A \backslash D_x$ and $y_\alpha = 0, \forall \alpha \in A \backslash D_y$. Consequently $x_\alpha . \overline{y_\alpha} = 0, \forall \alpha \in A \backslash D_x \cap D_y$ and the trivial inequality $|x_\alpha . \overline{y_\alpha}| \leq \frac{1}{2} \left(|x_\alpha|^2 + |y_\alpha|^2 \right)$ implies that the family $(x_\alpha . \overline{y_\alpha})$ is absolutely summable. Then we put $\langle x, y \rangle = \underset{\alpha \in D_x \cap D_y}{\sum} x . \overline{y_\alpha}$; it is easy to check that $\langle x, y \rangle$ is an inner product inducing the norm $\|x\|_2$, from which we deduce that $l_2(A)$ is a Hilbert space.

The outstanding fact about the spaces $l_2(A)$ is that they allow, in some sense, a classification of Hilbert spaces. More precisely we have:

Theorem 6.3.10. Let A be an arbitrary non empty set and let X be a Hilbert space, with an orthonormal base B having the same cardinal as A. Then there is an isometric isomorphism φ from X onto $l_2(A)$.

Proof: Since A and B have the same cardinal there is a bijection between them (see Chapter **1**, section **8**), and we can write $B = \{x_\alpha, \alpha \in A\}$. On the other hand let $x \in X$; according to Theorem **6.2.13** $(d), (f)$, we have $x = \sum_\alpha \langle x, x_\alpha \rangle . x_\alpha$ and $\|x\|^2 = \sum_\alpha |\langle x, x_\alpha \rangle|^2$. Therefore $(\langle x, x_\alpha \rangle, \alpha \in A) \in l_2(A)$ and we define $\varphi : X \longrightarrow l_2(A)$ by $\varphi(x) = (\langle x, x_\alpha \rangle, \alpha \in A)$ (Fourier coefficents of x). Theorem **6.2.13** (f) implies that φ is an isometry and since it is linear we deduce that it is injective. We prove that it is surjective. Let $a = (a_\alpha) \in l_2(A)$, then $\sum_\alpha |a_\alpha|^2 < \infty$; this implies that $a_\alpha = 0$ except for a sequence $\alpha_1, \alpha_2, \ldots$. Put $x = \sum_n a_{\alpha_n} . x_{\alpha_n}$;

apply the lemma below to get the convergence of this series, so x is well defined. Since $\{x_\alpha, \alpha \in A\}$ is orthonormal we have $a_\alpha = \langle x, x_\alpha \rangle, \forall \alpha \in A$ (why?), therefore $\varphi(x) = a$ and φ is surjective.■

Lemma. Let $y_1, y_2, \dots$ be an orthonormal sequence in a Hilbert space X and let $c_1, c_2, \dots$ be in $\mathbb{K}$. Then the series $\sum_n c_n.y_n$ converges iff $\sum_n |c_n|^2 < \infty$.

Proof: Consider the partial sums $\sum_{i=1}^{n} c_i.y_i$.

We have $\left\| \sum_{i=n}^{m} c_i.y_i \right\|^2 = \sum_{i=n}^{m} |c_i|^2 . \|y_i\|^2 = \sum_{i=n}^{m} |c_i|^2$ by the orthonormality of the y_n,

thus $\sum_{i=1}^{n} c_i.y_i$ is Cauchy in X iff $\sum_{1}^{m} |c_i|^2$ is Cauchy in $\mathbb{R}$.■

The classification of separable Hilbert spaces is given by:

Theorem 6.3.11. A Hilbert space X is separable if and only if X has a countable orthonormal base. If the base is finite the space is isometrically isomorphic to $\mathbb{K}^n$ for some n. If the base is infinite the space is isometrically isomorphic to $l_2(\mathbb{N})$.

Proof: Let B be an orthonormal base for X and let $x \neq y$ be in B. Then we have $(*)$ $\|x - y\|^2 = \|x\|^2 + \|y\|^2 = 2$. For each $x \in X$ consider the open ball $A_x = \left\{ y : \|x - y\| < \dfrac{1}{2} \right\}$ and observe that $t \neq s \implies A_t \cap A_s = \emptyset$, because of the relation $(*)$. Let D be a dense set in X. So each set A_x contains at least one point of D. Consequently, if B is not countable, the family $\{A_x, x \in B\}$ also will not be countable, so D will not be countable and X would not be separable.
Conversely suppose that B is countable, $B = \{x_1, x_2, \dots\}$. We know from Theorem **6.2.13** (c) that $\overline{M(B)} = X$ ($M(B)$ is the subspace generated by B). Now consider the family of vectors

$$D = \left\{ \sum_{1}^{n} a_k.x_k : n \geq 1, x_k \in B, a_k \in \mathbb{C}, a_k = s_k + i.t_k, \quad s_k, t_k \in \mathbb{Q} \right\}$$

It is not difficult to show that D is countable and that $M(B) \subset \overline{D}$; since $\overline{M(B)} = X$ we deduce that $\overline{D} = X$ and X is separable.■

4. Dual Space

Let X be a prehilbert space, we denote by X^* the topological dual of X, that is, the space of continuous linear functionals $\varphi : X \longrightarrow \mathbb{K}$, normed with $\|\varphi\| = \sup\{|\varphi(x)| : \|x\| \leq 1|\}$. We know that for every normed space X the dual X^* is a Banach space.
When X is a Hilbert space, we will see that X is isometric to its dual. Let us start with the following:

Theorem 6.4.1. Let X be a prehilbert space. For each $y \in X$, the function
$\quad \varphi_y : X \longrightarrow \mathbb{K}$ given by $\varphi_y(x) = \langle x, y \rangle$, $x \in X$
is a continuous linear functional with norm $\|\varphi_y\| = \|y\|$.

Proof: It is clear that φ_y is linear. By the Cauchy-Schwarz inequality, for every $x, y \in X$ we have: $|\langle x, y \rangle|^2 \leq \langle x, x \rangle . \langle y, y \rangle$. So $\|\varphi_y(x)\| \leq \|x\| \|y\|$. This proves that

φ_y is continuous and $\left\|\varphi_y\right\| \leq \|y\|$. On the other hand $\left|\varphi_y(y)\right| = |\langle y, y \rangle| = \|y\|^2 \leq \left\|\varphi_y\right\| \cdot \|y\|$, thus $\left\|\varphi_y\right\| \geq \|y\|$, whence $\left\|\varphi_y\right\| = \|y\|$.$\blacksquare$

Theorem 6.4.2. (Riesz)

Let X be a Hilbert space, with dual X^*. Then we have:

(a) For each $\varphi \in X^*$, there is a unique $y \in X$ such that $\varphi = \varphi_y$.

(b) The function $y \longrightarrow \varphi_y$ is a bijective isometry of X onto X^*.

Proof: (a) Let $\varphi \in X^*$, and let $N = \{x \in X : \varphi(x) = 0\}$ be the kernel of φ and suppose φ not identically null. Then $N \neq X$ and N is closed, so $X = N \oplus N^\perp$. We have $N^\perp \neq \{0\}$ and there is $y_0 \in N^\perp$, $y_0 \neq 0$. Let us observe that the two linear functions φ, φ_{y_0} are null on the same subspace, that is N, so there is $\lambda \in \mathbb{K}$ such that $\varphi = \lambda.\varphi_{y_0}$ (see Appendix **3**).We deduce that $\varphi(x) = \lambda.\varphi_{y_0}(x) = \lambda.\langle x, y_0 \rangle = \langle x, \overline{\lambda}.y_0 \rangle$, $\forall x \in X$. This gives $\varphi = \varphi_y$, with $y = \lambda.y_0$. The uniqueness is clear since if $\langle x, y \rangle = \langle x, y' \rangle$, $\forall x \in X$, we have $y - y' \perp x$, whence $y - y' = 0$.

(b) We have $\varphi_{y+y'} = \varphi_y + \varphi_{y'}$ and $\varphi_{\lambda y} = \overline{\lambda}.\varphi_y$, this proves that $y \longrightarrow \varphi_y$ is semilinear if $\mathbb{K} = \mathbb{C}$ and is linear if $\mathbb{K} = \mathbb{R}$. Since $\left\|\varphi_y\right\| = \|y\|$, it comes that $y \longrightarrow \varphi_y$ is one-to-one, and by (a) it is onto.$\blacksquare$

5. Exercises

117. Let X be a prehilbert space with inner product $\langle x, y \rangle$.

Prove that the function $(x, y) \longrightarrow \langle x, y \rangle$ from $X \times X$ into $\mathbb{K}$ is continuous.

118. Show that Cauchy-Schwarz inequality (Theorem **6.1.4**(a)) is strict iff the vectors x, y are linearly independent, which means: there is no scalar λ with $y = \lambda.x$

.

119. Let X be a prehilbert space and A a subset of X. Define:

$A^\perp = \{x \in X : x \perp a, \quad \forall a \in A\}$

(a) Prove that $A^\perp$ is a closed subspace of X.

(b) $A \subset B \implies B^\perp \subset A^\perp$

(c) $A \subset \left(A^\perp\right)^\perp$.

120. Let X be a Hilbert space and M is a closed subspace of X.

Prove that $M = \left(M^\perp\right)^\perp$.

121. Let M be a closed subspace of a Hilbert space X. and define the mapping $P : X \longrightarrow M$, by:

$x \in X$, $P(x) = y$, where y is the orthogonal projection of x on M.

Prove that P is linear, continuous and satisfies $P^2 = P \circ P = P$.

122. Let X be a Hilbert space and $P : X \longrightarrow X$, a linear continuous operator of X. Assume $P^2 = P \circ P = P$. and define the subsets of X:

$M = \{x \in X : P(x) = x\}, \quad N = \{x \in X : P(x) = 0\}$

(a) Prove that M and N are closed subspaces of X and $M \cap N = \{0\}$.

(b) Prove that vector $x \in X$ has a unique representation of the form:

$x = y + z$, with $y \in M$ and $z \in N$.

(c) Prove that $N = M^\perp$.

123. Let $x_1, x_2, ..., x_n$ be n linearly independent vectors in a Hilbert space X. Let M be the subspace generated by $x_1, x_2, ..., x_n$.

Prove that the orthogonal projection of the vector $x \in X$ on M is the vector $\sum_1^n \langle x, x_i \rangle . x_i$ (observe that M is a closed subspace of X and apply Lemma **6.2.3**).

124. (Gram-Schmidt Process)

Let $B = \{x_1, x_2, ...\}$ be a sequence of linearly independent vectors in a Hilbert space X. We define the sequence of vectors $e_1, e_2, ...$ by the process:

$$e_1 = \frac{x_1}{\|x_1\|}, \ e_2 = \frac{x_2 - \langle x_2, e_1 \rangle}{\|x_2 - \langle x_2, e_1 \rangle\|}, ..., e_n = \frac{x_n - \sum_1^{n-1} \langle x_n, e_i \rangle . e_i}{\left\| x_n - \sum_1^{n-1} \langle x_n, e_i \rangle . e_i \right\|}$$

Note that the vector $\sum_1^{n-1} \langle x_n, e_i \rangle . e_i$ is the orthogonal projection of the vector x_n on the subspace $M(e_1, e_2, ..., e_n)$ generated by $e_1, e_2, ..., e_n$ (Exercise **123**).

(a) Prove that $M(e_1, e_2, ..., e_{n-1}) = M(x_1, x_2, ..., x_{n-1})$, $\forall n$.

Deduce that $x_n \neq \sum_1^{n-1} \langle x_n, e_i \rangle . e_i$ and so e_n is well defined.

(b) Prove that the sequence $\{e_1, e_2, ..., e_n, ...\}$ is an orthonormal base of the closed subspace $\overline{M(B)}$.

125. Let $A : X \longrightarrow X$ be a linear continuous operator of a Hilbert space X.

(a) Show that for each $y \in X$ the function $x \longrightarrow \langle Ax, y \rangle$ is in the dual X^*. So there is a unique vector $\alpha_y \in X$ such that $\langle Ax, y \rangle = \langle x, \alpha_y \rangle, \forall x \in X$.

(b) Let $A^* : X \longrightarrow X$ be the mapping given by: $A^*(y) = \alpha_y$

Prove that A^* is a linear continuous operator of X with norm $\|A^*\| = \|A\|$.

(A^* is the adjoint operator of the operator A satisfying $\langle Ax, y \rangle = \langle x, A^*(y) \rangle$)

(c) Prove that:

(i) $(A \circ B)^* = B^* \circ A^*$

(ii) $A = A^*$ (A selfadjoint) $\Longleftrightarrow$ $\langle Ax, x \rangle$ is real $\forall x \in X$.

126. Let $A : X \longrightarrow X$ be a linear continuous operator of a Hilbert space X.

Prove that the following properties are equivalent:

(a) $A^* \circ A = I$, (I is the idendity operator of X)

(b) $\langle Ax, Ay \rangle = \langle x, y \rangle, \forall x, y \in X$

(c) $\|Ax\| = \|x\|, \forall x \in X$

CHAPTER 7

Topological Vector Spaces

This chapter may be considered as an introduction to topological vector spaces with some basic properties. The notion of completeness, we will deal with, is of particular importance. This will be done through the concept of generalized Cauchy sequences.

1. Compatible topology on a vector space

Definition 7.1.1. Let X be a vector space on the field $\mathbb{K}$ $(= \mathbb{R}$ or $\mathbb{C})$ and let τ be a topology on X. We say that τ is compatible with respect to the vector structure of X if the following mappings:

$(x, y) \longrightarrow x + y$, from $X \times X$ into X

$(\lambda, x) \longrightarrow \lambda.x$, from $\mathbb{K} \times X$ into X

are continuous with respect to the product topology on $X \times X$ and $\mathbb{K} \times X$.
The vector space X endowed with such topology is called a topological vector space, abreviated (t.v.s).

Examples 7.1.2. (a) Every normed space is a t.v.s.

(b) The space $\mathbb{K}^n$ with the product topology is a t.v.s.

Note the following consequence of the definition:

Theorem 7.1.3. In a t.v.s X:
The translations $\tau_a : x \longrightarrow x + a$, $a \in X$, and the homothetic map $\varphi_\lambda : x \longrightarrow \lambda.x$, $\lambda \in \mathbb{K}$, $\lambda \neq 0$ are homeomorphisms of X.

Proof: The inverse of the translation τ_a is the translation τ_{-a} and the inverse of the homothetic map φ_λ is the homothetic map $\varphi_{\lambda^{-1}}$; so it is enough to prove continuity. But it comes from Definition **7.1.1** and the following decompositions:

$\tau_a : x \longrightarrow (x, a) \longrightarrow x + a$

$\varphi_\lambda : x \longrightarrow (\lambda, x) \longrightarrow \lambda.x$

■

Corollary For $x \in X$, denote by $\mathcal{V}(x)$ the family of neighborhoods of x. Then:

$W \in \mathcal{V}(x) \iff \exists V \in \mathcal{V}(0) : W = x + V$

where $x + V = \{x + z, \ z \in V\}$.

Notations
Let X be a vector space on the field $\mathbb{K}$ and $A, B \subset X$, $\Lambda \subset \mathbb{K}$, we define:

$A + B = \{x + y, \ x \in A, y \in B\}$

$x + A = \{x + z, \ z \in A\}$

$\Lambda.A = \{\lambda.x : \lambda \in \Lambda, x \in A\}$

$-A = \{-x : x \in A\}$

Theorem 7.1.4. In a t.v.s X:

(a) The mapping $u : (x, y) \longrightarrow x - y$, is continuous from $X \times X$ into X

(b) For each neighborhood V in $\mathcal{V}(0)$ there is a W in $\mathcal{V}(0)$ such that
$$W + W \subset V$$

(c) For each neighborhood V in $\mathcal{V}(0)$ there is a W in $\mathcal{V}(0)$ such that
$$W - W \subset V$$

(d) For each neighborhood V in $\mathcal{V}(0)$ there is a W in $\mathcal{V}(0)$
and there is $\epsilon > 0$ such that
$$\Lambda_\epsilon.W \subset V, \text{ where } \Lambda_\epsilon = \{\lambda \in \mathbb{K} : |\lambda| < \epsilon\}$$

Proof: (a) Write u as $g \circ f$, where $f : X \times X \longrightarrow X \times X$ is given by $f(x, y) = (x, -y)$ and $g : X \times X \longrightarrow X$ defined by $g(x, y) = x + y$. Now f is continuous since its components $\pi_1(x, -y) = x$ and $\pi_2(x, -y) = -y$ are abviously continuous. The mapping is continuous by the definition of a t.v.s.

(b) comes from the continuity of $(x, y) \longrightarrow x + y$ at $(0, 0)$, indeed if V is a neighborhood of 0 in X there is W_1, W_2 in $\mathcal{V}(0)$ such that $W_1 + W_2 \subset V$, then $W = W_1 \cap W_2$ works.

(c) proof similar to that of (b).

(d) comes from the continuity of $(\lambda, x) \longrightarrow \lambda.x$ at $(0, 0)$. ∎

Definition 7.1.5. A subset A of a t.v.s is balanced if $\lambda.A \subset A$ for all $\lambda \in \mathbb{K}$, with $|\lambda| \leq 1$, that is A is invariant by the mappings $x \longrightarrow \lambda.x$ for all $\lambda \in \mathbb{K}$, with $|\lambda| \leq 1$.

Theorem 7.1.6. In a t.v.s X, the family of open balanced neighborhoods of 0 is a base for $\mathcal{V}(0)$. This means that for every $U \in \mathcal{V}(0)$ there is an open balanced neighborhood V of 0 with $V \subset U$.

Proof: By Theorem **7.1.4**(d) for each neighborhood U in $\mathcal{V}(0)$ there is an open set W in $\mathcal{V}(0)$ and there is $\epsilon > 0$ such that
$$\Lambda_\epsilon.W \subset U, \text{ where } \Lambda_\epsilon = \{\lambda \in \mathbb{K} : |\lambda| < \epsilon\}$$
On the other hand $\lambda.W$ is open (continuity of $x \longrightarrow \lambda.x$), so $\Lambda_\epsilon.W = \bigcup_{\lambda \in \Lambda_\epsilon} \lambda.W$ is open; moreover if $|\lambda| \leq 1$ we have $\lambda.\Lambda_\epsilon \subset \Lambda_\epsilon$ and then $\lambda.\Lambda_\epsilon.W \subset \Lambda_\epsilon.W$. This proves that the open set $V = \Lambda_\epsilon.W$ is balanced and, since $0 \in V \subset U$, achieves the proof. ∎

Definition 7.1.7. A t.v.s X is separated if it satisfies the axiom of Hausdorff: For all $x, y \in X$ with $x \neq y$, there is an open neighborhood $V(x)$ of x and an open neighborhood $V(y)$ of y such that $V(x) \cap V(y) = \emptyset$.

Theorem 7.1.8. A t.v.s X is separated if and only if for every $x \neq 0$ there is an open neighborhood V of 0 such that $x \notin V$.

Proof: The if part is clear. Conversely suppose the condition satisfied and let $x, y \in X$ with $x \neq y$. We have $z = x - y \neq 0$, then there is an open neighborhood V of 0 such that $z \notin V$. Let W be a neighborhood of 0 such that $W + W \subset V$ (Theorem **7.1.4**(b)). Put $U = W \cap -W$, $A = x + U$, $B = y + U$. Then U is a neighborhood of 0 with $U = -U$; A is a neighborhood of x; B is a neighborhood of y, and we have $A \cap B = \emptyset$, otherwise the vector $z = x - y$ would be in $U - U = U + U \subset W + W \subset V$, contradicting the choice of V. ∎

Theorem 7.1.9. A t.v.s X is separated if and only if $\{0\}$ is closed.

Proof: The if part is satisfied by Proposition **2.5.3**(c), Chapter **2**. Conversely if $\{0\}$ is closed, every $\{x\}$ is closed since $\{x\} = x + \{0\}$; so if $x \neq 0$, $0 \notin \{x\}$ and 0 is not a limit point of the closed set $\{x\}$, therefore there is a neighborhood V of 0 with $V \cap \{x\} = \emptyset$, i.e $x \notin V$. Then the conclusion comes from Theorem **7.1.8.**∎

2. Complete Topological Vector Spaces

The completeness property we give here for t.v.s is based on the concept of generalized Cauchy sequences. This approch is simple and avoids using uniform topological structures.

Definition 7.2.1. Let Λ be a set endowed with a partial ordering $\prec$. We say that Λ is a net for the relation $\prec$, if for every $\alpha, \beta \in \Lambda$ there is $\gamma \in \Lambda$ such that $\alpha \prec \gamma$ and $\beta \prec \gamma$.

Examples 7.2.2. (a) The sets $\mathbb{N}$ and $\mathbb{R}$ are nets for their usual ordering.

(b) The power set $\mathcal{P}(X)$ of any set X is a net for the inclusion partial ordering.

Definition 7.2.3. A generalized sequence in a set X is a mapping from a net Λ into X: $\alpha \in \Lambda \longrightarrow x_\alpha \in X$.
In the case $\Lambda = \mathbb{N}$ we get the usual sequence $x_n, n \in \mathbb{N}$.

Definition 7.2.4. Let X be a topological space and let $(x_\alpha)_{\alpha \in \Lambda}$ be a generalized sequence in X. We say that (x_α) converges to x if for each neighborhood V of x there is $\alpha_0 \in \Lambda$ such that $\forall \alpha \in \Lambda, \alpha_0 \prec \alpha \implies x_\alpha \in V$. Notation $\lim_\Lambda x_\alpha = x$.

Proposition 7.2.5. Let X be a separated topological space and let $(x_\alpha)_{\alpha \in \Lambda}$ be a generalized sequence in X. If (x_α) converges the limit of (x_α) is unique.

Proof: Similar to the usual sequences $x_n, n \in \mathbb{N}$.∎

Definition 7.2.6. Let X be a t.v.s and let $(x_\alpha)_{\alpha \in \Lambda}$ be a generalized sequence in X. We say that (x_α) is a generalized Cauchy sequence if for each neighborhood V of 0 there is $\alpha_0 \in \Lambda$ such that $\forall \alpha, \beta \in \Lambda, \ \alpha_0 \prec \alpha, \beta \implies x_\alpha - x_\beta \in V$.

Proposition 7.2.7. Let X be a t.v.s and let $(x_\alpha)_{\alpha \in \Lambda}$ be a convergente generalized sequence in X. Then (x_α) is a generalized Cauchy sequence.

Proof: Let x be the limit of (x_α). for each neighborhood U of 0 there is $\alpha_0 \in \Lambda$ such that $\forall \alpha \in \Lambda, \alpha_0 \prec \alpha \implies x_\alpha \in x + U$. By Theorem **7.1.4**$(c)$, if V is a neighborhood of 0 there is a neighborhood W of 0 such that $W - W \subset V$. Now there is $\alpha_0 \in \Lambda$ such that $\forall \alpha, \beta \in \Lambda, \alpha_0 \prec \alpha, \beta \implies x_\alpha \in x + W, x_\beta \in x + W$, but then $x_\alpha - x_\beta \in W - W \subset V$.∎

Definition 7.2.8. A t.v.s X is complete if every generalized Cauchy sequence in X converges in X. A subset E of X is complete if every generalized Cauchy sequence in E converges in E.

Theorem 7.2.9. Let A be a subset of a t.v.s X and let x be a limit point of A, then there is generalized sequence $(x_\alpha)_{\alpha \in \Lambda}$ in A such that $\lim_\Lambda x_\alpha = x$. Moreover the net Λ is independent of A and x, but depends only on X.

Proof: Put $\Lambda = \mathcal{B}(0)$, where $\mathcal{B}(0)$ is a base of neighborhoods of 0 and define on Λ the following partial ordering:

$$U, V \in \Lambda \quad U < V \iff U \supset V$$

It is clear that Λ is a net. For each $U \in \Lambda$, choose $x_U \in x + U \cap A$, this is possible since $x \in \overline{A}$. We prove that the generalized sequence $(x_U)_{U \in \Lambda}$ in A converges to x. Let W be a neighborhood of 0. There is $U \in \Lambda$ with $W \supset U$, because $\Lambda = \mathcal{B}(0)$. Now if $V \in \Lambda$ and $U \supset V$, we have $W \supset V$ and $x_V \in x + W$, which gives $\lim_{\Lambda} x_U = x$.$\blacksquare$

Theorem 7.2.10. (a) In a complete t.v.s every closed subset is complete.

(b) In a separated t.v.s every complete subset is closed.

Proof: (a) Let X be a complete t.v.s and let A be a closed subset in X. Let $(x_\alpha)_{\alpha \in \Lambda}$ be generalized Cauchy in A. Then (x_α) is Cauchy in X so x_α converges to x since X is complete. But $(x_\alpha) \subset A$ then x is a limit point of A and $x \in A$ because A is closed. So A is complete.

(b) Let X be a separated t.v.s and let A be a complete subset in X. We prove that A is closed. Let $x \in \overline{A}$, there is generalized sequence $(x_\alpha)_{\alpha \in \Lambda}$ in A converging to x (Theorem **7.2.9**). This implies that (x_α) is Cauchy in A and then converges to $a \in A$ since A is complete. So $x = a$ by the uniqueness of the limit, this gives $x \in A$ and A is closed.$\blacksquare$

Theorem 7.2.11. Let X, Y be two t.v.s on the same field and let $f : X \longrightarrow Y$ be a linear bijection bicontinuous from X onto Y.

Then a subset E of X is complete if and only if $f(E)$ is complete.

Proof: Let $(x_\alpha)_{\alpha \in \Lambda}$ be generalized sequence in E. Then (x_α) is Cauchy in E iff $y_\alpha = f(x_\alpha)$, $\lambda \in \Lambda$ is Cauchy in $f(E)$, because f is linear bijective. Now the conclusion comes from the definition of completeness and the fact that f is bicontinuous.$\blacksquare$

CHAPTER 8

Topological vector spaces of finite dimension

1. Tychonoff Theorem

Theorem 8.1.1. (**Tychonoff**) Let X be a separated t.v.s of finite dimension n on the field $\mathbb{K} = \mathbb{R}$ or $\mathbb{C}$. Then:

    (a) Every isomorphism of vector spaces $\varphi : \mathbb{K}^n \longrightarrow X$, is bicontinuous.

    (b) The topology of X can be defined by means of a norm.

    (c) The topology of X is locally compact.

Proof. See [1]

2. Locally compact Normed spaces, Riesz Theorem

By the theorem above, any separated t.v.s of finite dimension is locally compact. The following theorem gives the converse:

Theorem 8.2.1 (Riesz) Every locally compact separated t.v.s is of finite dimension.

Proof. For general t.v.s the proof uses the uniform structure of the space which has not been considered in this book. So we give the proof in the case of a normed space, which is in fact the original version of the theorem. Then let X be a locally compact normed space. Let $B = \{x \in X : \|x\| \leq 1\}$ be the closed unit ball. We can consider B as compact, after modifying it, if necessary by an homothetic transformation. So B can be covered by a finite set of open balls $B\left(x_i, \dfrac{1}{2}\right), i = 1, 2, ...n$, of radius $\dfrac{1}{2}$. Let M be the subspace generated by $x_1, x_2, ..., x_n$. M is closed by Tychonoff theorem. We prove that $M = X$. To this end suppose there is $x \in X \setminus M$, so we have $d(x, M) = \alpha > 0$, where d is the distance induced by the norm of X. Therefore there is $y \in M$ such that $\alpha \leq \|x - y\| \leq \dfrac{3\alpha}{2}$. Let $z = \dfrac{1}{\|x - y\|} . (x - y)$ then $z \in B$ and there is $1 \leq i \leq n$ such that $\|z - x_i\| \leq \dfrac{1}{2}$. On the other hand we have $y + \|x - y\| .z = y + \|x - y\| .x_i + \|x - y\| . (z - x_i)$; since $y + \|x - y\| .x_i \in M$ we deduce that $\dfrac{1}{2} \|x - y\| \geq \|x - y\| . \|z - x_i\| \geq \alpha$, and then $\|x - y\| \geq 2\alpha$, which is a contradiction with the choice of $y.\blacksquare$

91

CHAPTER 9

The space $C(X)$

Let X be a compact topological space and let $C(X)$ be the Banach space of all continuous functions $f : X \longrightarrow \mathbb{R}$ with the uniform norm:

$\|f\| = \sup\{\, |f(x)| : \ x \in X\}, f \in C(X)$

Some properties of the space $C(X)$ are related essentially to the compactness structure of the space X. We present in this Chapter two fundamental theorems frequently used in several applications:

The Stone-Weirstrass Theorem which gives an identification of the subsets $A \subset C(X)$ that are dense in $C(X)$. As a consequence, the theorem allows some interesting approximations of continuous functions.

The Arzela-Ascoli Theorem which, by means of equicontinuity, describes conditions for the compactness of some subsets in $C(X)$.

1. Stone-Weirstrass Theorem

First let us observe that $C(X)$ is an algebra with the operations: $f + g$ and $f.g$, for $f, g \in C(X)$. Next a subset $A \subset C(X)$ is a subalgebra if it satisfies:

$f, g \in A, a, b \in \mathbb{R} \implies af + bg \in A, f.g \in A$

The Stone-Weirstrass Theorem gives conditions under which a subalgebra $A \subset C(X)$ is dense in $C(X)$.

The theorem will be a consequence of the following four lemma.

Lemma 9.1.1. Consider the real function $f(x) = |x|$, $-1 \leq x \leq 1$. then:

There exist a sequence of polynomials p_n, without constant term, converging uniformly to f on $[-1, 1]$, i.e such that $\lim\limits_{n} \|p_n - f\| = 0$.

Lemma 9.1.2. Let $A \subset C(X)$ be a subset of $C(X)$ satifying:

$f, g \in A \implies \max(f, g) \in A$ and $\min(f, g) \in A$

Let $f \in C(X)$ be such that $\forall x \neq y$ in X there is a sequence $(f_n) \subset A$ with:

$f_n(x) \longrightarrow f(x)$ and $f_n(y) \longrightarrow f(y)$

Then $f \in \overline{A}$.

Lemma 9.1.3. Let $A \subset C(X)$ be a closed subalgebra of $C(X)$, then:

$f \in A \implies |f| \in A$ and $f, g \in A \implies \max(f, g) \in A$ and $\min(f, g) \in A$

Lemma 9.1.4. Let $A \subset C(X)$ be a subalgebra of $C(X)$ with the properties:

$(*)$ $\forall x \neq y$ in X, there is $f \in A$ such that $f(x) \neq f(y)$

$(**)$ $\forall x \in X$, there is $f \in A$ such that $f(x) \neq 0$

Then $\forall x \neq y$ in X, and $\forall a, b$ in $\mathbb{R}$, there is $f \in A$ such that $f(x) = a, f(y) = b$.

We prove lemma 9.1.**3** and lemma 9.1.**4**, leaving the proof of the remaining lemmas to the reader.

Proof of lemma 9.1.**3**: Let $g \in A$; if p is a polynomial without constant term, then $p \circ g \in A$ because A is an algebra. Now let $f \in A, f \neq 0$, and put $\|f\| = M$, $g(x) = \dfrac{f(x)}{M}$,we have $|g(x)| \leq 1$. If p_n is the sequence of polynomials of lemma 9.1.1 then $\lim\limits_{n} \|p_n \circ g - |g|\| = 0$. Since A is closed we get $|g| \in A$ and so $|f| \in A$. to finish the proof use the relations $\max(f,g) = \dfrac{1}{2}(f + g + |f - g|)$ and $\min(f,g) = \dfrac{1}{2}(f + g - |f - g|)$.∎

Proof of lemma 9.1.**4**: The conditions $(*), (**)$ imply the existence of g, h, k in A such that:
$$g(x) \neq g(y), h(x) \neq 0, k(y) \neq 0$$
then we put:
$$u = g.k - g(x).k, \quad v = g.h - g(y).h$$
we have $u, v \in A$ and $u(x) = v(y) = 0, u(y) \neq 0, v(x) \neq 0$; we deduce that the function $f(t) = a.\dfrac{v(t)}{v(x)} + b.\dfrac{u(t)}{u(y)}$ works.∎

Now we are in a position to state Stone-Weirstrass Theorem.

Theorem 9.1.5. (Stone-Weirstrass) Let $A \subset C(X)$ be a subalgebra of $C(X)$ with the properties:
$(*)$ $\forall x \neq y$ in X, there is $f \in A$ such that $f(x) \neq f(y)$
$(**)$ $\forall x \in X$, there is $f \in A$ such that $f(x) \neq 0$
Then A is dense in $C(X)$. In other words if $f \in C(X)$ there is a sequence f_n in A such that $f_n \longrightarrow f$ uniformly on X.
(An algebra $A \subset C(X)$ satisfying $(*)$ is said to separate the points of X).
(An algebra $A \subset C(X)$ containing the constants satisfies $(**)$).

Proof. It is not difficult to prove that $\overline{A}$ is a closed subalgebra of $C(X)$, so by lemma **3**, we have $f, g \in A \implies \max(f,g) \in A$ and $\min(f,g) \in A$.

Let $g \in C(X)$. If $x \neq y$ in X, and $g(x) = a, g(y) = b$ in $\mathbb{R}$, there is $f \in A \subset \overline{A}$ such that $f(x) = a, f(y) = b$. This shows that the function g satisfies the conditions of lemma **9.1.2.** which implies that $g \in \overline{A}$. Finally $\overline{A} = C(X)$.∎

Here is a complex version of Theorem **9.1.5.**, $C(X)$ being the Banach space of complex valued fonctions on X.

Theorem 9.1.6. Let $A \subset C(X)$ be a subalgebra of $C(X)$ with the properties:
$(*)$ $\forall x \neq y$ in X, there is $f \in A$ such that $f(x) \neq f(y)$
$(**)$ $\forall x \in X$, there is $f \in A$ such that $f(x) \neq 0$
$(***)$ $f \in A \implies \overline{f} \in A$ where $\overline{f}$ is the conjugate of f.
Then $\overline{A} = C(X)$.∎

As an application of theorem **5** we have:

Theorem 9.1.7. Every continuous function $f : [0,1] \longrightarrow \mathbb{R}$, is the uniform limit of a sequence of polynomials on $[0,1]$; in other words for each $\epsilon > 0$ there is a polynomial P such that $|P(x) - f(x)| < \epsilon, \forall x \in [0,1]$.

Proof. In the Banach space $C[0,1]$, consider the algebra A of polynomials on $[0,1]$; it is clear that A contains the constants and separates the points of $[0,1]$, so by Theorem **9.1.5.** we have $\overline{A} = C[0,1]$.∎

2. Arzela -Ascoli Theorem

In this section we assume that X is a compact metric space, with distance d. Let $A \subset C(X)$ be a subset of $C(X)$. We say that A is equicontinuous if:

$$\forall \epsilon > 0 \quad \exists \delta = \delta_\epsilon \text{ such that } x, y \in X, d(x,y) < \delta \implies |f(x) - f(y)| < \epsilon, \forall f \in A.$$

Note that δ depends only on ϵ, but neither on x, y nor f.

Examples:

(a) Every finite family $\{f_1, f_2, ..., f_n\}$ of functions in $C(X)$ is equicontinuous.

(b) Every sequence $\{f_n, n \geq 1\} \subset C(X)$ uniformly convergent on X
  is equicontinuous.

Theorem 9.2.1. (Arzela -Ascoli)

Let $A \subset C(X)$ be an equicontinuous family satisfying:

$$\sup \{|f(x)| : f \in A\} < \infty \quad \text{for each } x \in X$$

Then A is uniformly bounded, that is, $\sup \{\|f\| : f \in A\} < \infty$ and $\overline{A}$ is compact.

See [1] for the proof.∎

Bibliography

[1] **Berberian, S.K** : Lectures in Functional Analysis and Operator Theory
Springer 1974.

[2] **Dungundji, J:** Topology,   Allyn and Bacon, Boston 1966.

[3] **Hewitt, E and Stromberg, K:**   Real and Abstract Analysis
Springer 1966.

[4] **Kelley, J.L:** General Topology, Springer Verlag 1975.

[5] **Royden, H.L and Fitzpatrick, P.M:**   Real Analysis Mac Millan 2011

[6] **Rudin, W** : Principles of Mathematical Analysis
Wiley. 1966

[7] **Rudin, W** : Real and Complex Analysis,
Mac Craw Hill 1970.

[8] **Rudin, W** : Functional Analysis,   Mac Craw Hill 1973.

[9] **Yosida, K:** Functional Analysis,   Springer Verlag 1980.

[10] **Wilanski, A:** Modern Methods in Topological Vector Spaces
Mac Craw Hill 1978.

Index

In this index, numbering is as follows:
numbers I, II, ... refer to chapters
numbers 1, 2, ... refer to sections
numbers in brackets refer to exercises

A

B

C

D

Norm (of an operator) V, 3

O

Open set II, 1
Open mapping theorem V, 7
Ordering I, 4
Orthogonality VI, 2
Orthonormal base VI, 3

P

Parseval relation VI, 3
Perfect set III, 9 (58)
Parallelogram law VI, 1
Partition of unity IV, 9 (80)
Prehilbert space VI, 1
Product topology II,3
Projection (orthogonal) VI, 2
Pythagorean theorem VI, 2

Q

Quotient topology II, 3 (22)
Quotient norm V, 1

R

Relatively compact IV, 9
Riesz theorem VI, 4; A 2

S

Separated space II, 5
Separable space III, 7
Seminorm V, 1(84)
Sphere III, 1
Stone-Weirstrass B 1
Summability VI, 3
Support of a function IV, 9

T

Tietze theorem II,5
Topology II,1
Topology (weak) V, 5
Total (ordering) I, 4
Totally bounded set IV, 3
Tychonoff theorems IV, 7 and A 1

U

Uniform boundedness theorem V, 7
Uniformly continuous function III, 5

yes
I want morebooks!

Buy your books fast and straightforward online - at one of world's fastest growing online book stores! Environmentally sound due to Print-on-Demand technologies.

Buy your books online at
www.morebooks.shop

Kaufen Sie Ihre Bücher schnell und unkompliziert online – auf einer der am schnellsten wachsenden Buchhandelsplattformen weltweit! Dank Print-On-Demand umwelt- und ressourcenschonend produziert.

Bücher schneller online kaufen
www.morebooks.shop

KS OmniScriptum Publishing
Brivibas gatve 197
LV-1039 Riga, Latvia
Telefax: +371 686 204 55

info@omniscriptum.com
www.omniscriptum.com

Printed by Books on Demand GmbH, Norderstedt / Germany